AWESOME
Kitchen Science Experiments
for Kids

Awesome KITCHEN SCIENCE EXPERIMENTS for Kids

50 S T E A M
PROJECTS YOU CAN EAT!

DR. MEGAN OLIVIA HALL

Photography by Paige Green

ROCKRIDGE PRESS

For general information on our other products and services or to obtain technical support, please contact our Customer Care Department within the United States at (866) 744-2665, or outside the United States at (510) 253-0500.

Rockridge Press publishes its books in a variety of electronic and print formats. Some content that appears in print may not be available in electronic books, and vice versa.
Interior and Cover Designer: Francesca Pacchini
Art Producer: Sue Smith
Editors: Orli Zuravicky and Eliza Kirby
Production Editor: Chris Gage

Photography © 2019 Paige Green
Styling by Alysia Andriola
Author photo courtesy Rebecca Palmer

ISBN: Print 978-1-64152-621-0 | eBook 978-1-64152-622-7

R0

For Julie Doble,
who taught me everything
I know about making
science fun—and how to grill
a delicious panzanella.

CONTENTS

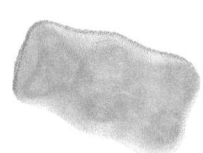

A NOTE FOR PARENTS

Awesome Kitchen Science Experiments for Kids is a window into the world of authentic—and delicious— STEAM experiments. Doing science experiments and cooking with your kids is special because it combines "Aha!" moments with yummy foods. Plus, in the comfort of your kitchen, your kids can practice essential science and engineering skills.

I wrote this book because I want every child to experience the fun of data-driven discovery and to connect their STEAM knowledge to everyday events. Too often, kids feel that they don't fit the STEAM mold. But every kid can ask questions, make observations, and develop explanations based on results gathered through hands-on learning. By doing so, your young kitchen scientists will experience the real STEAM deal and see themselves for the scientists and engineers that they already are.

At the top of each experiment is a note about how much adult prep is needed so you can plan ahead. Kitchen jobs that are too hazardous for kids are noted so you can chop vegetables or preheat the oven before your children come into the kitchen. Every experiment has a *Turn It Up!* variation for older kids. If you are working with younger children, read ahead before deciding if you want to suggest the advanced version. Please pay close attention to see if an experiment requires parental supervision, use of goggles, or other safety precautions. In general, any experiment involving chopping or cooking on the stove or in the oven is rated Medium or Advanced to show that parental supervision is needed.

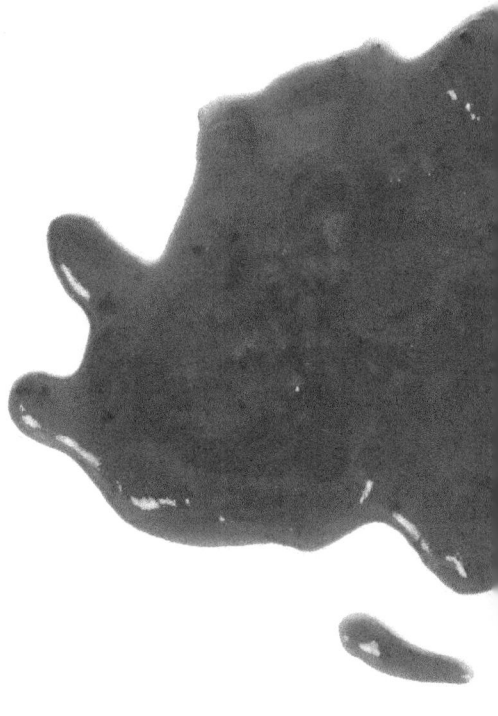

Glossary terms are bolded the first time they appear, but you and your child may not be doing the experiments in order. If your child sees a word they don't know, remind them to check the glossary before pulling out the dictionary—chances are it's in there!

Although many of the experiments can be done with little or no adult supervision, you may discover that you enjoy doing these experiments (and eating the results) with your kids. I've done most of them with my son Dylan, who is 10 years old. My 4-year-old daughter, Rosalea, has pitched in quite a few times. We have had a wonderful time cooking together, talking about what happened, and chowing down afterward. If you get as excited as I do about food, check out the Resources section for several fantastic adult reads on food science.

Science is messy and mistakes are to be expected. If the occasional experiment goes sideways, you are doing STEAM right. Don't expect every experiment to look pretty or turn out perfectly, but do expect to have fun. STEAM professionals know that experiment failure is not a scientist failure; it's a scientist win. Encountering the unexpected is how we make new discoveries. High-five your young food scientist the first time their experiment explodes (literally or figuratively)—and enjoy!

A NOTE FOR KIDS

My name is Dr. Megan Olivia Hall, but you can call me Megan. I've always loved food because I love to eat. I started cooking when I was seven years old. My specialties were macaroni and cheese and hamburgers. When I was a high school student, I cooked my family's meals and bought the groceries, too.

I discovered how much I loved science when I studied biology in college. The magic of biology led me to teaching science, and teaching got me hooked on technology, engineering, and mathematics. For the past 20 years, I've been lucky to work with thousands of amazingly curious kids in St. Paul Public Schools. In an awesome twist of fate, my years in the lab made me an even better cook.

Now I am excited to share what I know about kitchen science with you. In this book, you will find 50 edible experiments. Some experiments are wild and some are weird, but all of them can be tied to STEAM. I hope you have a blast with these 50 experiments, and that as you practice science in the kitchen, you realize what a fabulous scientist you are.

I have one last job for you. When you get cooking, invite other people to join you. You can invite friends, siblings, neighbors, or even (gasp!) your parents. STEAM challenges are meant to be shared, and most of these experiments make enough food for two to four people. Cooking and eating with friends and family are some of my most treasured memories, and they warm my heart like a fresh-from-the-oven cookie on a snowy day.

I hope these food experiments will warm your heart, too!

Chapter One

LET'S DIG IN

Science is about understanding the natural world by asking questions and finding answers. Food is a fantastic (and delicious) way to dig into those questions and explore **science**, **technology**, **engineering**, **arts**, and **mathematics**, also known as STEAM. Many types of science help explain how our food is cooked, what it tastes like, and even what it looks like. For example, scientists working in the field of **chemistry** study all the **matter** in the universe, explaining how different types of matter—like eggs and flour—change when they are heated, cooled, or combined with other substances. **Earth science** explains the minerals of our Earth—like table salt—and the **gases** in our atmosphere that make our meals fluffy or flat. Those are just a couple of ways STEAM helps us explore and understand food.

The cool food creations chefs make today are only possible because of advances in technology. Engineers use scientific knowledge and technology tools to find and solve problems, like how to keep egg whites fluffy in macaroon cookies. To build a gingerbread house, a chef needs to engineer a strong design and apply artful decorations.

None of this would happen without the number-crunching power of mathematics—which scientists, technicians, and engineers use every day—or without art, which brings beauty and meaning to our discoveries.

In this book, you will learn about food science, including the science behind cooking and baking. You'll make observations and ask questions. Before each experiment, you will guess what will happen—and soon you will have **results** to see what *really* happened. Best of all, at the end of each experiment, you will have something delicious to eat!

THE BIG DEAL ABOUT THE SCIENTIFIC METHOD

Scientists believe that everything we observe can be understood through careful study. Since the 1100s, countless scientists have developed the **scientific method** that STEAM professionals use today. That's what you're going to use in your experiments.

The scientific method is a series of steps.

1. **Scientists make observations.** Observations can be unexpected or surprising. It's often the strange things we observe that get us wondering about how the world works.
2. **Scientists ask questions.** A good scientific question is direct, specific, and answerable. For example, a good scientific question would be "Which ingredient in yeast bread makes the dough rise?"

3. **Scientists do background research.** If a scientific question has already been studied, good scientists will learn all they can about the discoveries from earlier studies.
4. **Scientists make educated guesses about the answers to their questions.** A **hypothesis** is a prediction about the answer to a scientific question.
5. **Scientists design experiments to test their hypotheses.** A great scientific experiment is focused on one idea. Scientists often design experiments that build on earlier experiments so that they can pick up where others have left off.
6. **Scientists analyze their experiment's results.** After scientists complete their experiments, they have to break down their results into understandable patterns. Analysis usually involves math.

Analyzing results will usually show that the hypothesis is supported or refuted (that is, shown to be false). It's very important that scientists use *only* the facts when analyzing their results, even if this shows something very different than what they expected. In science, honesty is more important than being right—and often, unexpected discoveries help the scientific community understand the world better.

You will be using the scientific method throughout this book. For every awesome kitchen experiment, there's a write-in box for your hypothesis, observations, and results. If you use a pencil, you have the option of erasing your notes when you come back to repeat an experiment.

THE KITCHEN IS A LAB

Science labs in movies are always full of people in white lab coats and shiny chrome surfaces. But if you look beneath the surface, the kitchen is like a lab—full of technology, tools, equipment, and supplies that you can use to explore answers to scientific questions—and a recipe is just like an experiment. To bring any recipe or experiment to life, you'll need tools. Let's take a look at the tools that will be used in this book.

KITCHEN APPLIANCES

- **Pots and pans:** You'll be using a variety of pots and pans to heat and cool your experiments.
- **Instant Pot®:** If you have an Instant Pot, you can use it in the yogurt-making experiment in chapter 3. Have an adult supervise your Instant Pot work.
- **Stove and oven:** Sometimes, you'll be heating your experiments. Always have an adult supervise any experiment involving the stove or oven.
- **Blender:** There's no faster way to chop up a **solid** than to pop it into a blender.
- **Silicone candy molds:** You can use a set of silicone candy molds to make foods into shapes, including the Real Rock Candy in this book.

SCIENTIFIC TOOLS

- **Measuring cups and spoons:** You'll use **liquid** and dry measuring cups as well as measuring spoons. A good set of measuring cups includes 1 cup, ½ cup, ⅓ cup, and

¼ cup. A good set of measuring spoons includes 1 tablespoon, 1 teaspoon, ½ teaspoon, ¼ teaspoon, and ⅛ teaspoon.

- **Containers:** Bowls, jars, bottles, cups, dishes, boxes, molds, and vases are all containers that will be used in these experiments.
- **Stirring implements:** In this book, you will use spoons, forks, mixers, whisks, spatulas, and your hands! (Don't forget to keep your hands clean in the kitchen.)
- **Cutting implements:** These include sharp knives, dull knives, and cookie cutters. Always have an adult present for any experiment that involves sharp tools.
- **Covers:** There will be times when you need reflective, airtight, or heat-proof layers on top of or underneath your experiments. Aluminum foil, plastic wrap, and parchment paper are key tools in the kitchen lab.
- **Matches:** Always have an adult present whenever you have an open flame in your kitchen lab.
- **Black light:** A small black light will help bioluminescent matter glow. Always shine lights away from your eyes.
- **Strainer:** Strainers separate solids from liquids and are very useful for phase-change experiments.
- **pH paper:** A pH indicator is available as a vial of tiny strips of paper that you can dip into your experiments to see how acidic or basic they are.
- **Candy thermometer:** Be sure to have a long, thin thermometer that you can lower into a soda can or clip on the side of a pot.
- **Weather thermometer:** You will need a thermometer that can measure ice-cold temperatures for making ice cream.

DOS AND DON'TS IN THE LAB

The kitchen lab is like any other lab: exciting, busy, and a little bit dangerous. Because all the experiments in this book are edible, they are also safe to touch, so you won't need to wear gloves—but you will need to keep your hands clean. Every time you begin work in the kitchen, wash the fronts and backs of your hands with hot, soapy water for as long as it takes to sing "Happy Birthday" twice. Don't forget your thumbs! Any time you touch something that isn't from the kitchen, including this book or even a cell phone, you will need to wash your hands again.

Be sure to have an adult supervise any experiments involving heat, sharp tools, or isopropyl alcohol. Be especially careful whenever using the stove or oven. Your adult helper should wear oven mitts when handling hot materials. You should both wear goggles to protect your eyes whenever glass, heat, sharp tools, or chemicals are present.

Chapter Two

HOW TO USE THIS BOOK

The experiments in this book are organized by STEAM subjects, with chapters for science, technology, engineering, art, and mathematics-based experiments. In each chapter, you'll find an explanation of the featured STEAM subject, a special feature on a cool topic in food science, and a series of fun—and edible!—experiments.

There is no right or wrong order in which to do the experiments in this book. Explore the different STEAM categories and experiments. As you read, you will see **bolded** vocabulary words that appear with definitions in the glossary. These words are bolded only the first time they appear. So if you're skipping around and see a word you don't know, check the glossary.

The most important thing to remember is to do the science that sparks your interest—whether you're choosing an experiment by STEAM subject, what you'd like to eat, or how fun the experiment looks.

GETTING READY

Each experiment or recipe follows a similar pattern. This section will show you exactly how to read each experiment and where to find information that might help you choose what experiment you want to do when.

Each experiment has a *Level of Difficulty*. *Easy* means you can do it on your own, although adults may need to prepare tools and ingredients. *Medium* means that you need an adult to supervise, but kids can mostly do it on their own. *Advanced* means you need an adult's help.

You will also find a *Mess-o-Meter* rating for each experiment. *Minor mess* means that the experiment will get your dishes and countertops sticky. *Medium mess* means that it might involve strange smells, items that need to be scraped away, and/or materials that need to sit out for multiple days.

In the *Best Eaten for* section, you'll see the meal for which this experiment is best eaten: breakfast, lunch, dinner, snack, or dessert. You will also see notes on the *Prep time* needed and the *Full time* needed to do the experiment. *Makes* will tell you how many servings of food your experiment will make.

Each experiment will also have a big question, marked with a ❓, you'll be trying to solve. Read the question and think for a moment before writing your hypothesis. Pay attention to the *Tools* and *Ingredients* sections. Make sure to read them carefully so you have everything you need before you begin.

Each experiment will also clearly call out any cautions, marked with a ⚠. Look here for any safety information such as the need for adult supervision or goggles.

Each experiment will have a write-in section for you to jot down notes during the experiment. You will write down your:

- *Hypothesis:* This is your predicted answer to the science question being asked.
- *Observations:* This is what you observe during the experiment.
- *Results:* This is the end product of the experiment and the answer to the science question being asked.

You will also see *The Steps* for conducting the experiment. After you've finished, check out *The Hows and Whys* section to understand the science behind the experiment. Then turn to the *STEAM Connection* to see what STEAM principles are present in the experiment. Finally, the *Turn It Up!* suggestions are ways to make the experiment more advanced for scientists who are ready for a challenge. If your experiment doesn't go as planned, that's okay! Even failed experiments are educational—and sometimes lead to exciting surprises.

Enjoy making delicious discoveries! The world of kitchen science is yours for the eating!

Chapter Three

SCIENCE

Let's get started with science. You'll investigate the science behind cooking by exploring a collection of food science mysteries. Science is detective work; it's about asking questions and searching for answers. In each of the following 15 experiments, you will investigate one scientific idea by observing bizarre and delicious food sensations.

You'll be doing some experiments that focus on bubbles—bubbles in fizzy drinks and the bubbles that make dough rise. You'll also separate and combine **mixtures** as you make butter and salad dressing. You will master **physical changes** by cooking eggs to make them solid and melting cheese for a gooey sandwich. This chapter also includes an experiment that involves building (and burning) edible candles. A scientist with a sweet tooth can have it all with strawberry jam, a rainbow of dried fruit, scones, and muffins. To wrap up this chapter, you can impress your family with homemade pizza—feel free to explain the science of gluten to them as they chow down.

Keep an eye out for chances to write down hypotheses, make observations, and analyze the results of your experiments. Every time you practice thinking like a scientist, your science skills grow. So why wait? To the kitchen lab!

FINDING YOUR FIZZ:
CARBONATION HACKS

LEVEL OF DIFFICULTY: EASY
MESS-O-METER RATING: MINOR MESS
BEST EATEN FOR: SNACK
PREP TIME: NONE
FULL TIME: 10 MINUTES
MAKES: 2 LEMONADE DRINKS

 When air bubbles through a drink, we call the fizz **carbonation**. In this experiment, you will add carbonation to lemonade using baking soda and dry ice. Will baking soda or dry ice create better fizzy lemonade?

Cautions: Never touch dry ice with your bare hands. Only adults should handle dry ice. Wear oven mitts or use a long-handled spoon.

TOOLS:

- 3 clear drinking glasses
- Measuring spoons
- A few dry ice pellets (carried by some Costco, Walmart, and Safeway stores; search online for a local dry ice supplier)

INGREDIENTS:

- 1 bottle of lemonade
- 1 teaspoon baking soda
- 1 teaspoon water

THE STEPS

1. Pour about 1 cup of lemonade into 1 clear glass.

2. Taste the lemonade in the first glass and record your observations.

3. Pour about 1 cup of lemonade into a second clear glass.

4. Add 1 teaspoon of baking soda to the second glass and stir.

5. Taste the lemonade in the second glass and record your observations.

CONTINUED

Hypothesis: Predict whether baking soda or dry ice will produce more delicious fizzy lemonade.

_____

_____

_____

_____

_____

Observations: What do you notice about the flavor and fizziness of each drink?

_____

_____

_____

_____

_____

Results: Compare the flavor and fizziness of each drink.

_____

_____

_____

_____

6. Have an adult add 3 to 5 pellets of dry ice to the third clear glass. Pour 1 teaspoon of water into the glass. This will stick the dry ice to the bottom of the glass.

7. Pour about 1 cup of lemonade into the third glass.

8. If the smoke is still bubbling, use a straw, scoop the lemonade out with a spoon, or wait until it stops. Taste the lemonade in the third glass and record your observations.

9. Compare the flavor and fizziness of each glass of lemonade. Which did you prefer? Record your results.

The Hows and Whys Carbonation bubbles are made of a gas called carbon dioxide. When you add baking soda to lemonade, a **chemical change** takes place: the **base** in baking soda combines with the **acid** in lemonade to make carbon dioxide bubbles. The lemonade is chemically changed, and it loses its flavor. When you add frozen carbon dioxide (dry ice) to lemonade, a physical change takes place: The solid carbon dioxide heats up quickly and turns into carbon dioxide gas bubbles. The lemonade stays acidic and tastes much better!

STEAM CONNECTION: Scientists study chemistry and physics to explore how materials change phases, from solid to liquid to gas and back again. Technology is needed in order to make dry ice, as it involves freezing carbon dioxide gas at very low temperatures.

Turn It Up! Pop Rocks® is a candy that fizzes in your mouth because it has carbon dioxide locked inside. A package of Pop Rocks has about 10 percent of the fizz of one fizzy drink. Try stirring Pop Rocks into a glass of lemonade. Do you notice more fizz?

KNEADING TO BREATHE:
YEAST BREAD

TOOLS:

- 2-cup liquid measuring cup, heat-proof
- Measuring cups and spoons
- Microwave oven
- Spoon
- Mug, heat-proof
- 2 mixing bowls
- Stand mixer with paddle and dough hook attachments (optional)
- Fork
- Clear plastic wrap
- Cutting board or clean counter
- Ruler
- Rolling pin
- Table knife
- Cookie sheet
- Parchment paper
- Oven
- Oven mitts

LEVEL OF DIFFICULTY: ADVANCED
MESS-O-METER RATING: MEDIUM MESS
BEST EATEN FOR: LUNCH OR DINNER
PREP TIME: 10 MINUTES TO GATHER INGREDIENTS
FULL TIME: 1 HOUR TO MAKE THE DOUGH, 2 HOURS (OR OVERNIGHT) FOR RISING, 30 MINUTES TO SHAPE THE ROLLS, 2 MORE HOURS (OR OVERNIGHT) FOR RISING, 15 MINUTES TO BAKE
MAKES: 12 DELICIOUS CRESCENT ROLLS

? When bakers make bread, they have the help of tiny creatures called yeast. In this lab, you will measure how high yeast makes bread dough rise. Why does yeast make dough rise?

! Cautions: Adult supervision is needed whenever you use the oven. If you are using an electric stand mixer, have an adult supervise your work.

THE STEPS

1. Pour 1 cup of milk into the 2-cup heat-proof liquid measuring cup. Microwave for 1 minute so that the milk is warm but not hot.

2. Add 2¼ teaspoons (1 packet) of active dry yeast to the warm milk. Stir with the spoon and set aside. (You may notice a few bubbles appear on the surface of the milk. This means that the yeast are starting their work!)

3. Put 4 tablespoons of butter in a heat-proof mug and microwave for 30 seconds so that it melts.

4. Measure 3 cups of flour, 1 teaspoon of salt, and ¼ cup of sugar into the mixing bowl and stir with the paddle attachment (or with a spoon) until combined.

5. Pour the yeast mixture and the melted butter into the flour mixture.

6. Crack the egg into the same mug that you used to melt the butter. Whisk the egg for a few moments with a fork to break up the yolk. Add to the mixing bowl.

7. Mix the flour, yeast, butter, and egg mixture with the paddle attachment (or a large spoon) until the dough sticks together in a messy ball.

8. Switch to the dough hook and mix for 6 minutes (or turn the dough onto a floured cutting board or countertop and knead it for 8 minutes).

9. Pour 1 teaspoon of oil into a second mixing bowl and use your hands to spread the oil until it coats the inside of the bowl.

10. Use a ruler to measure the height of your dough ball. Record your observations.

11. Set 1 tablespoon of butter out so that it warms up to room temperature.

12. Place your kneaded dough ball into the oiled bowl and cover the bowl with clear plastic wrap. Let the dough rise on the kitchen counter for 2 hours, or in the refrigerator overnight.

CONTINUED

INGREDIENTS:

- 1 cup milk
- 2¼ teaspoons (1 packet) active dry yeast
- 5 tablespoons unsalted butter, divided
- 3 cups flour, plus ¼ cup for rolling out the dough
- 1 teaspoon salt
- ¼ cup sugar
- 1 egg
- 1 teaspoon oil

Hypothesis: Estimate how many inches the bread dough will rise.

_____

_____

_____

_____

Observations: Observe how the dough changes when you leave it to rise.

_____

_____

_____

_____

Results: Did the yeast make the dough rise as many inches as you predicted?

_____

_____

_____

_____

13. Use a ruler to measure the height of your dough ball. Record your observations.

14. Place your risen dough on a floured cutting board or countertop. Use the rolling pin to roll it out into a circle. The dough should be between ¼-inch and ½-inch thick.

15. Use the table knife to spread 1 tablespoon of butter evenly over the dough circle.

16. Use the table knife to cut the dough circle into 12 wedges, as though you were cutting a pie into pieces.

17. Roll each dough piece into a crescent shape. Start on the wide edge and roll in.

18. Cover the cookie sheet with a piece of parchment paper.

19. Place the crescent rolls on the parchment paper. Cover the rolls loosely with clear plastic wrap. Leave them on the kitchen counter to double in size (about 2 hours).

20. Preheat the oven to 400°F. Bake the rolls for 12 to 15 minutes, until they are golden brown and a little bit crusty on top. Use the oven mitts when removing the pan of rolls from the oven. Let cool 5 to 10 minutes before eating. Enjoy!

The Hows and Whys Yeast are living creatures. In order to wake up, yeast need to eat a little bit of sugar. Like humans, yeast breathe in oxygen and breathe out carbon dioxide. The carbon dioxide that yeast breathe out creates bubbles in the dough, which makes it rise.

STEAM CONNECTION: Microbiology engineers have been perfecting countless varieties of yeast dough for thousands of years. Different types of yeast make different flavors of bread.

Turn It Up! Yeast love sugar. If you added more sugar to the warm milk that you used to wake up the yeast, would you get different results in your bread dough?

SWEET SCIENCE: STRAWBERRY JAM

LEVEL OF DIFFICULTY: MEDIUM
MESS-O-METER RATING: MINOR MESS
BEST EATEN FOR: BREAKFAST OR SNACK
PREP TIME: 10 MINUTES TO REMOVE STRAWBERRY LEAVES AND STEMS
FULL TIME: ABOUT 1 HOUR
MAKES: ABOUT 1 QUART OF JAM

? Food scientists use special chemicals called **thickening agents** to give jams and jellies the texture we love. Are there enough **natural** thickening agents in strawberries to make all-natural strawberry jam?

! **Cautions:** Adult supervision is needed when cooking on the stove top. Be careful to wear oven mitts and to stay clear of splatters. When making observations, look at the pot of jam from the side. Do not lean over the hot pot.

THE STEPS

1. Remove the stems and leaves from 1 quart of strawberries and put them in a large pot.

2. Use the potato masher to crush the strawberries in the pot. This will smell good!

TOOLS:

- 1 large pot with lid
- Measuring cups
- Potato masher
- 1 large spoon
- Stove
- Oven mitts
- 1 quart jar with lid

INGREDIENTS:

- 1 quart strawberries
- 3 cups sugar

Hypothesis: Predict whether your all-natural strawberry jam will thicken to a normal jam texture.

_____
_____
_____
_____
_____

CONTINUED

Observations: What changes do you observe while the jam cooks?

_____

_____

_____

_____

_____

_____

_____

_____

Results: How thick was your jam?

_____

_____

_____

_____

_____

_____

_____

_____

3. Add 3 cups of sugar to the pot and stir with a large spoon until everything is well combined.

4. Place the pot of strawberries and sugar on the stove and heat, stirring at least every 2 minutes, until the mixture starts to bubble and boil.

5. Once the mixture is boiling, stir it every minute for 15 to 25 minutes, until there is less liquid in the pot and the jam that cools on the spoon handle is thick enough so that it doesn't slide off. Do not boil your jam for more than 30 minutes even if you don't observe these changes.

6. Put the lid on the pot, remove the pot from heat, and let the jam cool for 10 minutes. While you are waiting, record your observations.

7. With adult supervision, carefully pour the hot jam into the quart jar and attach the lid.

8. Refrigerate the jam overnight.

9. Record your results.

The Hows and Whys Strawberries naturally have large amounts of a thickening agent called **pectin**. Pectin needs sugar (which you added to your jam) and acid (from the strawberries) to make the gel-like texture in jam. Strawberries have enough pectin that you don't need to add any extra thickening agents.

Turn It Up! Try making jam from other pectin-heavy fruits like quince, plums, gooseberries, pears, or apples. Oranges, which also have lots of pectin, are great for making marmalade.

AS THE WORLD CHURNS: FLAVORED BUTTERS

TOOLS:

- ⮕ Measuring cups
- ⮕ Clean plastic jar with a tight lid
- ⮕ 3 clean marbles
- ⮕ Strainer
- ⮕ Spoon

INGREDIENTS:

- ⮕ ½ cup heavy whipping cream
- ⮕ 1 teaspoon chopped fresh herbs or dried spices

Hypothesis: Predict what portion (¼, ½, or ¾) of cream is actually butter.

_____

_____

_____

_____

LEVEL OF DIFFICULTY: EASY

MESS-O-METER RATING: MINOR MESS

BEST EATEN FOR: BREAKFAST, LUNCH, OR DINNER

PREP TIME: NONE

FULL TIME: 15 MINUTES TO MAKE THE BUTTER, PLUS 10 MINUTES TO FLAVOR IT

MAKES: 2 TABLESPOONS BUTTER

 Butter is made by separating cream, which is a mixture of butter (fat) and buttermilk (liquid). How much of cream is butter, and how much is buttermilk?

THE STEPS

1. Pour ½ cup of heavy cream into a clean plastic jar.

2. Add 3 clean marbles to the jar.

3. Place the lid on the jar and close it tightly.

4. Shake the jar until a lump of butter floats at the top of the buttermilk (5 to 10 minutes). Take turns shaking the jar if your arms get tired. Record your observations.

5. Place a strainer over a bowl.

6. Empty the jar into the strainer. The liquid that goes into the bowl under the strainer is buttermilk, which is delicious in pancakes and muffins.

7. Take the butter out of the strainer and return it to the jar. Record your results.

8. With a spoon, gently mix 1 teaspoon of chopped fresh herbs (like chives and parsley) or dried spices (like cinnamon and nutmeg). For more flavored butter recipes, visit the BritCo flavored butter website listed in the Resources section.

9. Enjoy your flavored butter! What other flavor combinations can you imagine?

> **The Hows and Whys** When you shake cream, parts of the mixture begin to separate. The fat parts stick to each other and turn into butter. You can see how much of cream is butter and how much is buttermilk by comparing the amounts of butter and buttermilk at the end of your experiment.

STEAM CONNECTION: Chemical engineers often work with mixtures. In food labs, mixtures include salads, trail mix, and spaghetti. Chemical engineers also mix up medicines, soaps, glues, and more.

Turn It Up! Apply your math skills by measuring the **volume** of butter and buttermilk at the end of your experiment to answer the big question with numbers. How could you use a liquid measuring cup to measure both volumes?

Observations: What do you notice while you are shaking the cream?

_____
_____
_____
_____
_____
_____
_____
_____
_____

Results: About how much of your cream was actually butter?

_____
_____
_____
_____
_____
_____
_____
_____
_____
_____

SALAD SOLUTION:
DRESSING UP A MIXTURE

LEVEL OF DIFFICULTY: EASY
MESS-O-METER RATING: MINOR MESS
BEST EATEN FOR: LUNCH, DINNER, OR SNACK
PREP TIME: NONE
FULL TIME: 20 MINUTES
MAKES: 1½ CUPS SALAD DRESSING

TOOLS:

- 2-cup liquid measuring cup
- 2-cup (1 pint) jar with tight-fitting lid

INGREDIENTS:

- ¾ cup canola or vegetable oil
- ¼ cup vinegar
- ¼ cup maple syrup
- ¼ cup mustard

Hypothesis: Predict which ingredients will form a solution and which ingredients will stay separate.

_____

_____

_____

_____

Salad dressing is a **solution**, or a combination of ingredients that can't be separated. Salad dressing is *also* a mixture, because some of the ingredients can separate. In this experiment, you will create a delicious salad dressing and find out which ingredients will separate from your salad dressing: oil, vinegar, maple syrup, or mustard?

THE STEPS

1. Pour ¾ cup of canola or vegetable oil into a 2-cup liquid measuring cup.

2. Pour ¼ cup of vinegar on top of the oil to bring the liquid in the measuring cup to a total of 1 cup.

3. Pour ¼ cup of maple syrup to bring the liquid in the measuring cup to a total of 1¼ cups.

4. Add ¼ cup of mustard to bring the liquid in the measuring cup to a total of 1½ cups.

5. Record your observations.

6. Pour everything from the liquid measuring cup into the jar.

7. Place the lid on the jar and close it tightly.

8. Shake the jar for 30 seconds to thoroughly mix the salad dressing.

9. Set the jar down on the counter. Wait 10 minutes for the mixture to settle. Record your results.

10. Enjoy your salad dressing on top of your favorite salad!

The Hows and Whys Oil-based liquids and water-based liquids are difficult to mix. For this reason, the oil in your salad dressing will always separate from the vinegar, mustard, and maple syrup—no matter how much you shake it up.

STEAM CONNECTION: Mixtures and solutions are at the heart of the work in a chemistry lab. Scientists who work with liquids must know whether chemicals are truly combined or if they will separate. Chemists can separate solutions with technologies like chromatography, which you will learn about in the next chapter.

Turn It Up! Experiment with different ingredient combinations for salad dressings. Beginning with a base of ¾ cup of canola or vegetable oil and ¼ cup of vinegar, try adding lemon or lime juice, fresh herbs, and your favorite savory spices. Smell each herb and spice before you add it to decide what will taste good together.

Observations: As you pour each ingredient into the measuring cup, what happens?

_____
_____
_____
_____
_____

Results: Note which ingredients separated out of the salad dressing after you mixed it.

_____
_____
_____
_____
_____

COMPARING APPLES & ORANGES: FRUIT DENSITIES

LEVEL OF DIFFICULTY: EASY
MESS-O-METER RATING: MINOR MESS
BEST EATEN FOR: BREAKFAST, SNACK, OR DESSERT
PREP TIME: NONE
FULL TIME: 10 MINUTES
MAKES: 2 PIECES OF FRUIT

? Objects sink or float because of their **densities**. Density is a measurement of how much **mass**, or stuff, is packed into a certain amount of volume, or space. In this experiment, you will compare the densities of several different fruits by observing whether they float or sink in water. Are apples or oranges more dense?

THE STEPS

1. Fill a clear pitcher or vase ¾ full with water.

2. Gently place the apple in the water and record your observations.

3. Remove the apple and gently place the unpeeled orange in the water. Record your observations.

4. Remove the orange and peel it. Remove as much of the white **pith** as you can. Place the peeled orange in the water. Record your observations.

TOOLS:

- 1 large, clear pitcher or vase

INGREDIENTS:

- Water
- 1 apple
- 1 orange with its peel on
- Optional: 2 other small fruits, such as a plum and a grape

Hypothesis: Predict whether apples or oranges are more dense (more likely to sink in water).

_____
_____
_____
_____
_____

CONTINUED

Observations: Note which fruit(s) floated and which fruit(s) sank.

_____
_____
_____
_____
_____
_____
_____
_____
_____

Results: Which fruit was denser, the apple or the orange? How do you know?

_____
_____
_____
_____
_____
_____
_____
_____

5. Try placing a variety of other fruits in the water to compare their densities as well.

6. Record your results.

The Hows and Whys Apples are more dense than oranges because oranges have thick, fluffy peels. The peel of an orange makes it float in water.

STEAM CONNECTION: Engineers apply knowledge of density when building and designing boats, ships, and airplanes. Density is an important concept for mechanical engineers working on plumbing design problems. Environmental scientists study the density of oil and water when cleaning up oil spills.

Turn It Up! You can use mathematics to calculate actual density. Use a kitchen scale to measure the mass of each piece of fruit. To measure the volume of each piece of fruit, place the fruit into a liquid measuring cup with some water. Note how much the water volume increases after the fruit is added. To calculate the density of the fruit, divide its mass by its volume.

DRY IT YOURSELF:
NATURE'S CANDY

LEVEL OF DIFFICULTY: MEDIUM
MESS-O-METER RATING. MINOR MESS
BEST EATEN FOR: SNACK OR DESSERT
PREP TIME: 5 MINUTES TO SLICE FRUIT
FULL TIME: 10 MINUTES TO PREPARE THE FRUIT,
PLUS 3 TO 8 HOURS DRYING TIME
MAKES: AT LEAST 2 CUPS OF DRIED FRUIT

 Fresh, in-season fruit has a taste that can't be beat. That taste can be concentrated in do-it-yourself dried fruit. In this experiment, you will measure the drying times for three different fruits to compare the amount of water in each. Which do you think has the most water: plums, apples, or strawberries?

Cautions: Adult supervision is needed whenever you use the oven.

THE STEPS

1. Have an adult prepare your fruits by

   a. Washing them with water,
   b. Removing leaves, stems, and pits, and
   c. Slicing them into ¼-inch-thick pieces.

2. Preheat your oven to 170°F.

TOOLS:

- Parchment paper
- 3 cookie sheets
- Oven
- Oven mitts
- Fork
- Plastic or glass food storage containers

INGREDIENTS:

- 6 plums
- 2 apples
- 1 quart strawberries

Hypothesis: Predict which fruit will have the most water, the least water, and a medium amount of water.

_____

_____

_____

_____

_____

CONTINUED

Observations: Note the time it took for each type of fruit to dry.

_____

_____

_____

_____

_____

_____

_____

Results: Which fruit had the most water? How do you know?

_____

_____

_____

_____

_____

_____

_____

3. Line 3 cookie sheets with parchment paper.

4. Place the plums on one cookie sheet, the apples on another, and the strawberries on the third cookie sheet. Spread out the pieces so that they are not touching.

5. Place the cookie sheets in the oven.

6. Every 30 minutes, check your experiment to see if the fruit is dry. Every 2 hours, remove the pans from the oven and use a fork to flip the fruit pieces. It will take 3 to 8 hours for your fruit to dry.

7. When the fruit looks leathery, take it out of the oven. Record your observations.

8. Place the dried fruit in an open food storage container on the kitchen counter for 1 week. Every day, shake the container.

9. Dried fruit can be stored in the refrigerator for up to 2 months.

The Hows and Whys Fruit is mostly water. When you dry fruit, you remove the water, leaving behind extra-sweet, extra-flavorful bites of tasty goodness. The fruit that takes the longest to dry had the most water.

STEAM CONNECTION: Food engineers work to solve problems, like how to create lightweight foods that won't spoil for travelers, campers, hikers, and astronauts. Freeze-dry technology involves placing thinly sliced fruits in a freezer instead of an oven. The frozen water **sublimates**, going from solid to gas state without being a liquid at all.

Turn It Up! Try freeze-drying fruit. After your fruit is clean and sliced, place it on a mesh rack in the freezer. The fruit will freeze solid in about 1 hour and should be dry after about 1 week.

WHAT NUT?*
COMBUSTION REACTIONS YOU CAN EAT

LEVEL OF DIFFICULTY: MEDIUM
MESS-O-METER RATING: MEDIUM MESS
BEST EATEN FOR: SNACK
PREP TIME: 5 MINUTES TO PREPARE THE NUTS AND POTATO
FULL TIME: 15 MINUTES
MAKES: 1 NUTTY SNACK

? **Combustion** reactions involve a fuel and oxygen reacting to produce heat and light. All fires are combustion reactions. In this experiment, you will create fires using nut oil as your fuel. What nut burns the best: an almond, a peanut, or a pecan?

! **Cautions:** Have an adult prepare the nuts and potato. Adult supervision is needed while you work with an open flame.

THE STEPS

1. Cut 3 peanuts and 3 pecans into long, thin slivers.

2. Cut a potato into a long cylinder or tall rectangular cube.

CONTINUED

* For anyone with nut allergies, this experiment can be done using sunflower or pumpkin seeds.

TOOLS:

- A knife for an adult to prepare the nuts
- Matches
- Heat-proof surface (plate, pan, etc.)

INGREDIENTS:

- 3 peanuts
- 3 pecans
- 1 potato
- 3 slivered almonds

Hypothesis: Which nut will burn best: an almond, a peanut, or a pecan?

_____

_____

_____

_____

Observations: Note how easily each nut caught on fire and write down how many seconds each fire lasted.

_____

_____

_____

_____

_____

_____

Results: Which nut burned best? Use your observations to explain.

_____

_____

_____

_____

_____

_____

3. Stick 1 almond sliver into the top of the potato. The potato will hold the almond while it is burning.

4. Light the almond sliver on fire with a match. Let it burn until it goes out. Repeat with the other 2 almond slivers. Record your observations.

5. Repeat steps 3 and 4 with the peanut and pecan slivers. Record your observations and results.

6. You can eat the nuts you didn't burn.

The Hows and Whys The oil in nuts is a fuel that can be lit to create a fire. The oil and oxygen make a chemical change during the fire, turning into carbon dioxide, gas, and steam. The nut that lights the fastest and burns the longest has the most oil.

STEAM CONNECTION: Engineers study combustion reactions to design **engines** that burn fuel. Technologies that help make engines more **efficient** reduce the amount of carbon dioxide these reactions add to Earth's atmosphere. Hybrid car engines are one example of a fuel-efficient technology.

Turn It Up! When the nut slivers were burning on top of the potato, your experiment may have looked like a candle. This is a cool optical illusion.

SAY CHEESE: GETTING THE PERFECT GRILL

LEVEL OF DIFFICULTY: MEDIUM
MESS-O-METER RATING: MINOR MESS
BEST EATEN FOR: LUNCH OR DINNER
PREP TIME: 5 MINUTES TO SLICE THE CHEESE
FULL TIME: 15 MINUTES
MAKES: 1 GRILLED CHEESE SANDWICH

 A **phase change** is when solids melt, liquids evaporate or freeze, and gases condense. One of the most delicious phase changes in the kitchen is the melting of cheese. In this experiment, you will take a scientific approach to preparing a grilled cheese sandwich. At what temperature does cheese melt?

Cautions: Adult supervision is needed while cooking on the stove top.

THE STEPS

1. Spread 1 slice of bread with ½ tablespoon of butter. Place the bread in a cold frying pan, buttered side down.

2. Cover the bread with a single layer of cheese.

3. Spread the other slice of bread with ½ tablespoon of butter. Place the bread on top of the cheese, buttered side up.

4. Carefully insert the thermometer inside the sandwich.

5. Place the pan on the stove top over low heat.

CONTINUED

TOOLS:

- Table knife
- Frying pan
- Candy thermometer
- Stove
- Oven mitts
- Spatula

INGREDIENTS:

- 2 pieces of bread
- 1 tablespoon of butter
- Enough slices of cheese to cover one side of bread

Hypothesis: Predict what temperature will melt cheese.

_____

_____

_____

_____

_____

6. Observe the sandwich for 5 minutes, checking the temperature and state of the cheese. If the cheese is still solid after 5 minutes, turn the burner to medium.

7. Use the spatula to flip over the sandwich so that the bread does not burn.

8. Continue observing the temperature and state of the cheese. If the cheese is still solid after 5 more minutes, turn the burner to high.

9. When the cheese melts, turn off the burner and record your results.

The Hows and Whys When cheese melts, two different phase changes happen. The fat in the cheese melts first. The **proteins** melt second. Soft cheeses melt at lower temperatures and hard cheeses need hotter temperatures. Most cheeses melt between 130°F and 180°F.

STEAM CONNECTION: Scientists, technicians, and engineers need to understand the materials in their labs. The temperatures at which phase changes occur are important **physical properties**. These temperatures are listed for each element in the **periodic table**, the ultimate reference chart for chemists.

Turn It Up! Try toasting cheese. Cut a cube of cheese and fry it in the pan. You may need some oil to keep it from sticking. Keep the temperature low enough so that it doesn't melt.

Observations: Note the starting temperature. What happens with your sandwich as you slowly increase the temperature?

_____
_____
_____
_____
_____
_____
_____
_____
_____

Results: At what temperature did the cheese in your sandwich melt?

_____
_____
_____
_____
_____
_____
_____
_____
_____

GET SOME CULTURE:
HOMEGROWN YOGURT

TOOLS:

- Large pot
- Stove
- Oven mitts
- Measuring cups
- Candy thermometer
- Spoon
- Yogurt maker (available online for about $30) or Instant Pot®

INGREDIENTS:

- ½ gallon milk
- ½ cup yogurt (containing active cultures)

Hypothesis: Predict how the yogurt bacteria will change the milk.

_____

_____

_____

_____

_____

LEVEL OF DIFFICULTY: MEDIUM

MESS-O-METER RATING: MINOR MESS

BEST EATEN FOR: BREAKFAST OR SNACK

PREP TIME: NONE

FULL TIME: 45 MINUTES TO MAKE THE CULTURE AND 3 TO 12 HOURS FOR YOGURT INCUBATION

MAKES: 2 QUARTS OF YOGURT

? Yogurt is a **culture**, or growth, of tiny creatures called **bacteria**. In this lab, you will grow a culture of yogurt bacteria by adding a small amount of active yogurt bacteria to a large amount of milk. What will the yogurt bacteria do with the milk?

! Cautions: Adult supervision is needed while you are cooking on the stove. Use a machine designed to make yogurt (either a yogurt maker or an Instant Pot®) for this experiment. Other methods may make a yogurt that is not safe to eat.

THE STEPS

1. Set a large pot on the stove and pour in ½ gallon of milk.

2. Turn the stove burner to a medium setting and begin stirring the milk.

3. Heat the milk to 180°F, stirring constantly.

4. Turn off the burner. Keep the thermometer in the milk. Let the milk sit until it cools to 115°F.

5. Stir ½ cup of yogurt into the milk.

6. Pour the milk and yogurt mixture into the jars of your yogurt maker (or into the pot of your Instant Pot®) and incubate according to the manufacturer's directions. It is very important that your yogurt stays between 105°F and 112°F for the entire incubation period.

7. Record your observations and results.

Observations: Record the look, smell, and taste of your yogurt culture when it is finished.

_____
_____
_____
_____
_____

The Hows and Whys Warm milk is perfect for yogurt bacteria. The milk is full of sugars that yogurt bacteria love to eat. Although bacteria are very small, within a few hours the yogurt bacteria multiplied so much that it ate the milk and turned it into yogurt.

Results: Note how the yogurt is different from the milk.

_____
_____
_____
_____
_____

STEAM CONNECTION: Yogurt bacteria are an important part of the culture of billions of bacteria living in human intestines. These gut flora help us digest our food and ease stomach pain. Yogurt is one of many technologies that doctors use to help people who have problems with their gut flora.

Turn It Up! Try making yogurt with a different kind of milk, or starter yogurt. How does your new yogurt taste different? Why do you think that is?

MAKING A MUFFIN OUT OF A MOLEHILL:
SHAPE-CHANGING SCONES

LEVEL OF DIFFICULTY: ADVANCED

MESS-O-METER RATING: MEDIUM MESS

BEST EATEN FOR: BREAKFAST OR SNACK

PREP TIME: 5 MINUTES TO GATHER INGREDIENTS

FULL TIME: 30 MINUTES TO MIX, PLUS 30 MINUTES COMBINED BAKING TIME

MAKES: 12 SCONES AND 12 MUFFINS

TOOLS:

- Oven
- Oven mitts
- 2 large bowls
- Large spoon
- Measuring cups and spoons
- 4-cup liquid measuring cup
- Cutting board or countertop
- Table knife
- Cookie sheet
- Pastry brush
- Small bowl
- Small spoon
- Fork
- Muffin tin
- 12 paper muffin tin liners
- Toothpick

 Many baked treats are soft and fluffy because of **leavening**, an ingredient that causes dough to rise. In this experiment, you will explore the rising power of baking powder. What will baking powder do to raise scones? What will baking powder do to raise muffins?

Cautions: Adult supervision is needed when using the oven.

THE STEPS

1. Preheat the oven to 400°F.

2. In a large bowl, combine 4 cups of flour, 2 tablespoons of baking powder, 1 teaspoon of salt, and ⅔ cup of sugar. Use the spoon to stir the dry ingredients.

3. Stir in 1 cup of dried fruit.

CONTINUED

INGREDIENTS:

- 4 cups flour, plus ¼ cup for rolling
- 2 tablespoons baking powder
- 1 teaspoon salt
- ⅔ cup sugar, plus ¼ cup
- 1 cup dried fruit, small (dried currants) or diced (apricots, apples, pineapple)
- 2½ cups heavy whipping cream
- 1 teaspoon cinnamon
- ½ cup milk
- 2 eggs
- ½ cup chopped fresh fruit

Hypothesis: What differences do you think you will observe in how scone and muffin doughs rise in the oven?

_____

_____

_____

_____

4. Stir in 2½ cups of heavy cream until the dough is evenly mixed and sticks together in clumps.

5. Scoop half of the dough into the second large bowl.

6. With clean hands, work the dough in the first bowl into a ball. Squish the dough together several times, until it sticks together. This is your scone dough. Record your observations.

7. Sprinkle ¼ cup flour onto a cutting board.

8. Move the dough ball out of the first bowl and onto the floured surface.

9. With your hands, shape the dough into a long rectangle, about ¾ inch tall and 3 inches wide. Cut the rectangle into 6 squares. Cut each square into 2 triangles.

10. Move the triangles onto a cookie sheet. Dip the pastry brush into the cream that is left in the measuring cup. Brush the triangles with the cream.

11. In a small bowl, mix ¼ cup of sugar with 1 teaspoon of cinnamon. Using a small spoon, sprinkle the tops of the scones with about half of this mixture. Save the rest.

12. Place the scones in the oven and bake 12 to 15 minutes, until the tops are light golden brown and firm to the touch.

13. While the scones are baking, measure ½ cup of milk in the liquid measuring cup. Crack 2 eggs into the milk. Stir this mixture with a fork and add it to the dough in the second bowl along with ½ cup of chopped fresh fruit. Stir everything well with the large spoon. This is your muffin dough. Record your observations.

14. Place the muffin tin liners in the muffin tin.

15. Using your large spoon, fill each muffin tin liner ²/₃ full with the dough from the second bowl.

16. Sprinkle the rest of your cinnamon-sugar mixture over the tops of the muffins.

17. When the scones are done baking, remove them from the oven and turn the oven down to 375°F. Bake the muffins for 15 to 20 minutes, until a toothpick inserted into a muffin comes out clean.

18. Sample the scones and the muffins. Record your results.

The Hows and Whys Baking powder causes dough to rise by creating a chemical change that makes bubbles. Double-acting baking powder creates two chemical changes: one change when it reacts with wet ingredients, and a second change when the dough gets hot. Muffins have wetter, lighter batter than scones, so baking powder makes them rise taller when they bake.

STEAM CONNECTION: In commercial food labs, engineers work to find the perfect amount of baking powder for products such as cake mix, corn muffins, and brownies. Can you imagine working in a lab where brownies are always baking?

Turn It Up! Change the amount of baking powder or cream in the recipe to change the texture. If you want to experiment with flavors, try other types of fruit.

Observations: Note any similarities and differences between the scone dough and the muffin dough.

_____
_____
_____
_____
_____
_____
_____
_____

Results: Compare the tastes and textures of the scones and muffins.

_____
_____
_____
_____
_____
_____
_____
_____
_____

VEGETABLE PROTECTION:
WHAT STOPS ONION TEARS?

TOOLS:

- Knife for slicing onion
- Cutting board
- Small frying pan
- Stove
- Plate
- Fork

INGREDIENTS:

- 1 onion (preferably older)
- 1 teaspoon olive oil

Hypothesis: Share an idea for preventing onion tears and say why you think it will work.

_____
_____
_____
_____
_____

LEVEL OF DIFFICULTY: MEDIUM
MESS-O-METER RATING: MEDIUM MESS
BEST EATEN FOR: SNACK
PREP TIME: 15 MINUTES FOR INTERVIEWS
FULL TIME: 20 MINUTES
MAKES: ABOUT ½ CUP CARAMELIZED ONIONS

 Many people burst into tears when slicing up an onion. In this experiment, you will test different ways to protect yourself from this. What do you think will prevent onion tears?

Cautions: Have an adult slice the onion. Adult supervision is needed when using the stove.

THE STEPS

1. Ask at least 3 different people if they have any ideas for how to stop an onion from making you cry when you slice it. Here are some additional ideas:

   a. Freeze the onion.
   b. Hold something (like a matchstick or a piece of bread) in your mouth while you are slicing the onion.
   c. Wear goggles while you slice the onion.

2. Choose at least 2 ideas to try.

3. Stand over the cutting board while an adult slices your onion very thinly. Test one idea while the adult slices half

the onion, and then switch to the second idea while the adult slices the second half. Record your observations.

4. Heat 1 teaspoon of olive oil in a small frying pan over high heat.

5. Add the onion and cook for 3 to 6 minutes, until the edges of the onion slices start to turn brown.

6. Turn the burner down low. Cook the onion on low heat until it is soft and caramel brown.

7. Turn off the stove. Move the onion, which is now caramelized (cooked to release sugars) onto a plate. Taste the onion. Are you crying? Did cooking the onion fix the onion tears problem? Record your results.

The Hows and Whys When you cut an onion, it releases a gas that is irritating to your eyes. This gas becomes more concentrated over time. Very freshly picked onions will not make you cry, but older onions sure will. Cooking the onion will break down the gas. Scientists don't yet know how to stop eye irritation with raw onions.

STEAM CONNECTION: For many years, scientists did not know why slicing onions makes people cry. In 2002, scientists in Japan discovered the irritating gas. Now that we know what causes onion tears, perhaps engineers will create a technology that prevents them.

Turn It Up! Try caramelizing carrots, sweet peppers, and celery. How do the vegetables change?

Observations: Note what happened when you tried to prevent onion tears.

_____

_____

_____

_____

_____

_____

_____

_____

Results: Which ideas worked? Which didn't?

_____

_____

_____

_____

_____

_____

_____

IF YOU CAN'T TAKE THE HEAT: CHLOROPHYLL ON BOIL

TOOLS:

- Small pot
- Stove
- Timer
- Large bowl
- Slotted spoon
- Plate

INGREDIENTS:

- 1 cup pea pods
- 6 cups water, divided
- 12 ice cubes

Hypothesis: Predict whether blanching and cold shocking will make pea pods brighter or duller green.

_____

_____

_____

_____

_____

LEVEL OF DIFFICULTY: MEDIUM
MESS-O-METER RATING: MINOR MESS
BEST EATEN FOR: SNACK
PREP TIME: 5 MINUTES
FULL TIME: 15 MINUTES
MAKES: 1 CUP OF CRISP AND TENDER PEA PODS

 Cooked vegetables are more tender than raw ones, but can get mushy. In this experiment, you will try **blanching** and **cold shocking**, two techniques that together make vegetables crisp yet tender. How will the vegetables' color change?

Cautions: Adult supervision is needed whenever you are cooking on the stove.

THE STEPS

1. Use your hands to snap off the stems of 1 cup of pea pods.

2. Set 1 pea pod aside as your first **control**.

3. Pour 3 cups of water into a small pot.

4. Set the pot over a stove burner on high.

5. While the water comes to a boil, put 12 ice cubes in a bowl. Add 3 cups of cold water to the ice.

6. When the water is boiling, put the pea pods (except the one you set aside) into the water.

7. Set the timer for 2 minutes.

8. After 2 minutes, use the slotted spoon to transfer the pea pods into the bowl of ice water. Leave 1 pea pod in the hot water as your second control. Record your observations.

9. Set the timer for 5 minutes.

10. After 5 minutes, use the slotted spoon to transfer the pea pods in the bowl of ice water onto a plate.

11. Use the slotted spoon to transfer the pea pod in the hot water onto the edge of the same plate. Place your raw pea pod onto the other edge of the plate.

12. Record your observations and results.

Observations: Note the color changes throughout your experiment.

———————————————
———————————————
———————————————
———————————————

Results: What color and texture changes did you see?

———————————————
———————————————
———————————————
———————————————

The Hows and Whys When you boiled your pea pods, the heat caused some of the air in the plant to evaporate. When the air was gone, the green color, a **molecule** called **chlorophyll**, showed up more clearly. When green vegetables boil for a long time, the heat breaks down the chlorophyll, making them dull and mushy. The ice water stops the heat and keeps colors bright.

STEAM CONNECTION: Temperature has a big impact on living molecules, so biologists are especially aware of temperature.

Turn It Up! Try this technique with more green vegetables. Do you think it could work with vegetables of other colors? Give it a try!

HUMPTY DUMPTY: WHEN PROTEINS UNRAVEL

LEVEL OF DIFFICULTY: MEDIUM
MESS-O-METER RATING: MINOR MESS
BEST EATEN FOR: BREAKFAST
PREP TIME: NONE
FULL TIME: 10 MINUTES
MAKES: 1 FRIED EGG

TOOLS:

- Frying pan
- Stove
- Spatula
- Plate
- Fork

INGREDIENTS:

- 1 teaspoon butter or olive oil
- 1 egg
- Salt to taste

Hypothesis: Predict how you think an egg will change as you cook it.

_____
_____
_____
_____
_____

When eggs cook, they change in ways that can't be reversed. In this experiment, you will observe changes in an egg as it cooks on the stove. How does cooking change an egg?

Cautions: Adult supervision is required when cooking on the stove. If you've never cracked an egg before, ask for adult help with this step, too. Do not eat raw or undercooked eggs.

THE STEPS

1. Put 1 teaspoon of butter or olive oil in a small frying pan on the stove.

2. Turn the burner to medium heat.

3. Crack open 1 egg and drop its insides into the pan.

4. Cook the egg until it changes color. Record your observations.

5. Use the spatula to flip over the egg. Record your observations.

6. When both sides of the egg are white and the middle is firm, turn off the stove.

7. Put the egg on a plate and sprinkle it with salt. Taste the egg. Record your results.

The Hows and Whys Eggs are made of molecules called proteins. Proteins change shape when they are heated. If proteins are heated above 108°F, they **denature**—they change shape and can't change back to normal. Egg whites are 100 percent protein, so you can easily see egg proteins denature when you fry an egg.

STEAM CONNECTION: Until recently, scientists believed that protein denaturing was permanent. But in 2015, engineers in California discovered that you could un-boil an egg. They used chemical and machine technologies to help denatured egg proteins find their shape again.

Turn It Up! Try these different ways to cook eggs. Soft-boil an egg by lowering it into boiling water for 3 minutes or hard-boil it for 14 minutes. Crack an egg into a pot of boiling water and let it cook for 3 minutes to get a poached egg. Beat an egg with a fork and 1 teaspoon of water and fry this mixture in butter for fluffy scrambled eggs.

Observations: Note the colors and textures of a cooking egg.

_____

_____

_____

_____

_____

_____

Results: How did your egg change when you cooked it?

_____

_____

_____

_____

_____

_____

_____

I'M A GLUTEN FOR PIZZA: THE FIELD OF FLOURS

TOOLS:

- Measuring cups and spoons
- 2-cup liquid measuring cup, heat-proof
- Microwave oven
- 2 large mixing bowls
- Electric stand mixer (or large spoon)
- Cutting board or countertop
- Kitchen towel
- Rolling pin
- Ruler
- Pastry brush
- Oven
- Oven mitts

LEVEL OF DIFFICULTY: ADVANCED
MESS-O-METER RATING: MINOR MESS
BEST EATEN FOR: LUNCH OR DINNER
PREP TIME: NONE
FULL TIME: 30 MINUTES TO MAKE DOUGH, 1 HOUR FOR THE DOUGH TO RISE, 45 MINUTES TO ASSEMBLE AND BAKE PIZZAS
MAKES: 2 LARGE PIZZAS OR 4 PERSONAL-SIZE PIZZAS

? Gluten is a molecule that holds dough together. Some types of flour have more gluten than others. In this experiment, you will compare the amount of gluten in white and whole wheat flours. Will white or whole wheat flour produce stretchier dough and pizza crust?

! **Cautions:** Adult supervision is needed whenever you use the oven. If you are using an electric mixer, have an adult supervise your work.

THE STEPS

1. Measure ¾ cup of water in a 2-cup heat-proof liquid measuring cup. Heat in the microwave for 45 seconds so that the water is warm but not hot.

2. Stir 1½ teaspoons of yeast and 1 teaspoon of honey into the water. Set aside.

3. In a large mixing bowl, combine 2½ cups of white flour, ½ teaspoon of salt, and 1½ tablespoons of olive oil. Add the yeast mixture.

4. Mix with the paddle attachment on an electric stand mixer (or stir with a spoon) for 2 minutes.

5. Switch to the dough hook attachment (or turn the dough out onto a lightly floured cutting board or countertop). Mix for 10 minutes (or knead gently by hand for 10 minutes).

6. Shape the dough into a round ball. Grasp half of the ball in each hand, with your thumbs touching in the middle. Gently pull the dough in half. Stop when the dough starts to tear and measure how far you were able to stretch the dough. Record your observations. Roll the dough back into a ball and place it in the second large bowl.

7. Repeat steps 1 to 6, using 2¼ cups of whole wheat flour instead of white flour. Record your observations.

8. Place a clean kitchen towel over the bowl with both dough balls in it. Leave the bowl on the kitchen counter to rise for 1 hour.

9. Preheat the oven to 425°F.

10. If you are making 2 large pizzas, keep the dough in two large balls. If you are making 4 personal-size pizzas, divide each dough ball in half.

11. Use the rolling pin to flatten each pizza dough ball into a circle. The dough should be ⅜-inch thick. Place each pizza crust on a floured cookie sheet.

CONTINUED

INGREDIENTS:

- 1½ cups water, divided
- 3 teaspoons yeast, divided
- 1 teaspoon honey
- 2½ cups white flour, plus ¼ cup for rolling out the dough
- 1 teaspoon salt, divided
- 5 tablespoons olive oil, divided
- 2¼ cups whole wheat flour
- Pizza toppings: ½ cup red marinara sauce (such as canned spaghetti sauce), 16 ounces grated mozzarella cheese, and any veggies or meats you enjoy on pizza

Hypothesis: Predict which pizza crust will have more gluten and be stretchier: white pizza crust or whole wheat pizza crust.

_____

_____

_____

_____

Observations: Note the texture and stretchiness of the two doughs, as well as the taste and texture of the two crusts.

_____

_____

_____

_____

_____

Results: Which type of crust was stretchier? Which crust had more gluten?

_____

_____

_____

_____

_____

12. Brush 1 tablespoon of olive oil on each large pizza crust (½ tablespoon for personal-size pizzas).

13. Spread ⅛ to ¼ cup of red sauce on top of each pizza crust. Top with shredded mozzarella cheese and any other pizza toppings you enjoy.

14. Bake the pizzas for 15 to 20 minutes, until the crust is brown and crisp.

15. After you eat your pizza, record your observations and results.

The Hows and Whys Gluten is a sticky, stretchy molecule that works by trapping the air bubbles made by yeast, giving pizza crust its delicious texture. There is more gluten in white flour than in whole wheat flour, so white flour makes stretchier dough. Gluten is found in wheat, barley, rye, and oats, but not in corn or rice.

STEAM CONNECTION: Not everyone can eat gluten. Some flours have more gluten than others, and some have none at all. Engineers work to develop gluten-free products that have similar texture and taste but can be enjoyed by those with gluten allergies or intolerances.

Turn It Up! Try this recipe with corn or rice flour. How is your gluten-free pizza crust the same and different?

EXTRA FEATURE: SUPERMARKET SCIENCE

Science is all around you—even inside the grocery store!

FRUITS & VEGGIES

The first rule of shopping for fruits and vegetables is to buy in-season produce. Many fruits and vegetables taste better—and are better for you—at certain times of the year. When you buy in-season fruits and vegetables, they are more likely to have come from a nearby farm, have spent less time on a truck, and have more active vitamins and nutrients. Check the color of your produce—bright colors are good. Give each item a sniff. If it smells good, it's ripe. If it doesn't smell, it's not ripe yet. If it smells bad, do not buy it. Never buy fruits or vegetables from a package or bin with mold that you can see.

CHECKING DATES

Most packaged foods have a date on them. These dates will tell you that the food is "best by" a certain day, that grocers should "sell by" a certain day, or that the food will "expire" on a certain day. Expiration means that the food is not safe to eat anymore. "Best by" dates tell us when the food will stop tasting good.

NUTRITION LABELS

Packaged foods also have nutrition labels. In general, picking a food with less sugar and more protein, fiber, and vitamins will give you better energy and make you feel healthier. Nutrition labels will also tell you how many **calories** are in the food. High-calorie foods have more energy in them.

Chapter Four

TECHNOLOGY

Whenever someone applies science to solve a real-world problem, they're using technology. We often forget that everyday tools were once brand new. Imagine the problems that were solved when Europeans started using a new technology called the *fork* in the 1600s. Measuring cups, spatulas, whisks, blenders, and microwave ovens are all examples of kitchen technologies that solved real problems for cooks.

In this chapter, you'll be using technologies from culinary, **biology**, chemistry, and physics labs. You will learn how to make your food glow, how to turn liquid beverages into solid foods, and how to isolate **DNA** and plant **pigments**. You'll also build a simple version of a machine called a **calorimeter**, an oven that measures the amount of energy in a given food.

Because the chapter 4 experiments involve advanced technologies, they are the most advanced in this book. You may need a little extra parental supervision to complete each experiment safely, but don't let that stop you! These technologies are totally worth the time—you will be amazed at what you can do in your home kitchen.

LET IT GLOW:
BIOLUMINESCENCE IN YOUR KITCHEN

TOOLS:

- 1 (4-cup) liquid measuring cup, clear and heat-proof
- Microwave oven
- Spoon
- 4 clear 8-ounce cups (for the Jell-O)
- Small black light or black flashlight

INGREDIENTS:

- 4 cups tonic water, divided
- 1 (3-ounce) package of Jell-O
- 2 cups water, divided
- 1 (3-ounce) package of another flavor of Jell-O

LEVEL OF DIFFICULTY: MEDIUM
MESS-O-METER RATING: MINOR MESS
BEST EATEN FOR: SNACK OR DESSERT
PREP TIME: NONE
FULL TIME: 10 MINUTES TO MAKE, 4 HOURS FOR JELL-O TO SET IN THE REFRIGERATOR
MAKES: 4 SERVINGS

 **Bioluminescence** is the scientific word for living things that glow, including some jellyfish and deep-sea creatures. In this lab, you will be experimenting to discover which flavor (and color) of Jell-O, when made with a special kind of water, glows brightest. Do different flavors of Jell-O glow differently?

Cautions: Hot liquids can be dangerous. Have an adult pour the boiling tonic water and hot Jell-O.

THE STEPS

1. Pour 1 cup of tonic water into the clear, heat-proof 4-cup liquid measuring cup.

2. Place the measuring cup with the tonic water in the microwave and heat for 1 minute.

3. Check to see if the tonic water is boiling. If so, move on to step 4. If not, continue microwaving the water in 30-second stages until it boils, and then move on to step 4.

4. Pour one package of Jell-O into the hot tonic water and stir until it dissolves.

5. Add 1 cup of cold or room-temperature tonic water to the Jell-O solution and stir it for 10 seconds.

6. Pour the Jell-O solution into 2 of your 4 clear, 8-ounce cups.

7. Repeat steps 1 to 6 with the second package of Jell-O.

CONTINUED

Hypothesis: What flavor of Jell-O will glow brightest under the black light? Why?

_____

_____

_____

_____

_____

Observations: Describe what you see when you shine the black light through each color of Jell-O.

_____

_____

_____

_____

_____

Results: What color of Jell-O glowed brightest under the black light?

_____

_____

_____

_____

8. Put the Jell-O in the refrigerator to chill until it sets (2 to 4 hours).

9. Shine the black light through each color of Jell-O. What do you see? Record your observations and results.

The Hows and Whys Tonic water is a slightly bitter soda made with quinine. Quinine glows under a black light. The quinine reflects ultraviolet light from the black light as a visible, glowing light. The ultraviolet light looks different when reflected through different colors of Jell-O.

STEAM CONNECTION: Bioluminescence has inspired awesome scientific discoveries in biology labs, where scientists stick tiny glowing chemicals onto brain cells, viruses, antibodies, and DNA.

Turn It Up! Although your bioluminescent Jell-O is edible, you may have noticed a bitter taste. Tonic water has a bitter flavor. How could you alter this experiment to make glowing Jell-O that tastes better?

UNMASKING THE CODE OF LIFE: DNA EXTRACTION SMOOTHIES

LEVEL OF DIFFICULTY: ADVANCED

MESS-O-METER RATING: MINOR MESS

BEST EATEN FOR: BREAKFAST OR SNACK

PREP TIME: CHILL ISOPROPYL ALCOHOL IN THE REFRIGERATOR FOR AT LEAST 1 HOUR BEFORE BEGINNING THE EXPERIMENT

FULL TIME: 30 MINUTES

MAKES: 2 SMOOTHIES

Scientists often need samples of DNA, a special molecule that is different for each and every **organism**, to solve mysteries and answer questions. All cells contain DNA, but it can be tricky to get the DNA out. What tools do scientists need to take DNA out of cells?

Cautions: Always check that the blender lid is secure before plugging in the blender. Isopropyl alcohol can sting if it splashes in your eyes. Wear goggles and have an adult pour the isopropyl alcohol.

THE STEPS

1. In the blender or food processor, blend 1 banana with 1 cup of water until smooth.

CONTINUED

TOOLS:

- Blender or food processor
- Measuring cups and spoons
- 2 small (4- to 6-ounce) plastic cups or containers
- Spoon
- #2 cone coffee filter
- Rubber band
- 2 large drinking glasses

INGREDIENTS:

- 2 bananas
- 1 cup water, plus 4 teaspoons
- 1 teaspoon clear-colored shampoo
- 2 pinches table salt
- 1 cup frozen strawberries or raspberries
- 1 cup orange juice
- 4 teaspoons cold isopropyl alcohol (rubbing alcohol)

Hypothesis: Which ingredients from the list will pull DNA out of cells?

_____

_____

_____

_____

Observations: Throughout the experiment, watch for DNA to appear as a white cloud.

_____

_____

_____

_____

_____

Results: At which stage in your experiment did the DNA appear? What ingredients did the trick?

_____

_____

_____

_____

_____

2. In 1 of the 2 small plastic cups, SLOWLY stir 1 teaspoon of shampoo, 2 pinches of salt, and 4 teaspoons of water. Do not let this mixture foam.

3. Add 3 teaspoons of the banana-water mixture from step 1 to the shampoo-salt-water mixture from step 2. Stir slowly for 5 to 10 minutes.

4. Place the #2 cone coffee filter inside the second small plastic cup. Open the filter and fold the filter's wide-open edge over the lip of the cup. Make sure that the bottom of the filter is not touching the bottom of the cup. Secure the filter with the rubber band.

5. Pour the mixture from step 3 into the filter from step 4. A clear liquid will slowly cover the bottom of the filter cup.

6. As your mixture filters, go back to your blender. Place 1 banana, 1 cup of frozen strawberries, and 1 cup of orange juice in the blender. Blend until smooth. Pour the smoothies into 2 drinking glasses and enjoy.

7. By now, you most likely have at least 1 teaspoon of clear liquid at the bottom of your filter cup. Carefully pull off the rubber band and lift up the filter without disturbing the clear liquid. Throw away the rubber band and filter.

8. Take your isopropyl alcohol out of the refrigerator. Very slowly, pour a layer of cold isopropyl alcohol on top of the clear liquid. Let the solution sit for 2 to 3 minutes. The banana's DNA will appear, looking like a white cloud floating in the clear liquid.

The Hows and Whys Three ingredients helped **extract** DNA in this experiment: salt, shampoo, and isopropyl alcohol. The shampoo dissolved the molecules covering the DNA deep within each banana cell. The salt helped the tiny DNA molecules stick to each other. The isopropyl alcohol made the DNA visible.

STEAM CONNECTION: **Biotechnology** is a growing field in which scientists and engineers use technology to understand living things. DNA contains the code of life, and DNA extraction allows scientists to understand this code.

Turn It Up! All living cells contain DNA. Try this procedure with strawberries or homegrown wheat grass.

MARVELOUS MARBLES:
MOLECULAR GASTRONOMY FOR THE YOUNG CHEF

LEVEL OF DIFFICULTY: MEDIUM

MESS-O-METER RATING: MINOR MESS

BEST EATEN FOR: SNACK, OR AS A **GARNISH** WITH PANCAKES

PREP TIME: CHILL 2 CUPS OIL FOR 4 HOURS BEFORE BEGINNING THE EXPERIMENT

FULL TIME: 30 MINUTES TO MAKE THE MARBLES AND 5 MINUTES FOR THE TASTE TEST

MAKES: ABOUT ½ CUP OF MARBLES

 **Spherification** is a **culinary** technology in which chefs turn liquids into flavorful solid marbles. Any liquid—juice, hot chocolate, even soup—can be turned into tiny bites of goodness. Does changing the texture of a familiar food change the taste?

Cautions: Hot liquids can be dangerous. Have an adult pour the steaming juice.

THE STEPS

1. Pour 2 tablespoons of juice into 1 of the 2 liquid measuring cups.

2. Pour 1 envelope of gelatin over the juice and stir for 20 seconds.

3. Pour ¼ cup of juice into the second liquid measuring cup.

TOOLS:

- Measuring spoons
- 2 (2-cup) liquid measuring cups, clear and heat-proof
- Spoon
- Microwave oven
- Squeeze bottle
- Strainer
- Bowl

INGREDIENTS:

- 2 tablespoons juice, plus ¼ cup
- 1 (¼-ounce) envelope of unflavored gelatin
- 2 cups cold canola or vegetable oil

CONTINUED

Hypothesis: Will your experimental subject be able to identify the flavor of your marvelous marbles?

_____

_____

_____

_____

Observations: What flavors did your experimental subject guess?

_____

_____

_____

_____

Results: Was your experimental subject able to guess the flavor?

_____

_____

_____

_____

4. Microwave for 10 seconds, and then stir.

5. Repeat until the juice is steaming but not boiling.

6. Pour the steaming juice into the liquid measuring cup from steps 1 and 2 and stir for 1 minute.

7. Pour the juice-gelatin mixture from step 6 into the squeeze bottle.

8. Chill the squeeze bottle in the refrigerator for 10 minutes but no longer (otherwise it will solidify).

9. Take the squeeze bottle and the cold oil out of the refrigerator.

10. Slowly drip the juice-gelatin mixture from the squeeze bottle into the cold oil. Play with the size of the drops you make—try a variety!

11. When you have used up all of your juice-gelatin mixture, slowly pour the cold oil and the juice-gelatin marbles into your strainer. You can put a container under the strainer if you want to use the oil for another batch of marbles.

12. When the oil is gone, rinse any remaining oil off the marbles.

13. Pour the marbles from the strainer into a bowl.

14. Find an experimental subject (like a parent or a sibling) to taste a marble and tell you what it is made out of. Could they recognize the flavor, even though the juice was in solid form? Record your observations and results.

The Hows and Whys When gelatin gets cold, it goes from liquid to solid. In this experiment, you got your gelatin cold so quickly that it formed separate marbles. The marbles didn't stick to each other because gelatin can't dissolve in oil. But the flavor of the juice didn't change.

STEAM CONNECTION: Spherification is part of a larger culinary field called molecular gastronomy in which chefs use chemistry to separate out the molecules of flavor and nutrition that make food fabulous.

Turn It Up! For no-mess pancakes, make marbles out of maple syrup. Prepare a recipe of pancakes, sprinkling 3 to 7 maple syrup marbles on top of each pancake just before you flip it. When you bite down into the pancakes, the marbles will release the maple syrup flavor.

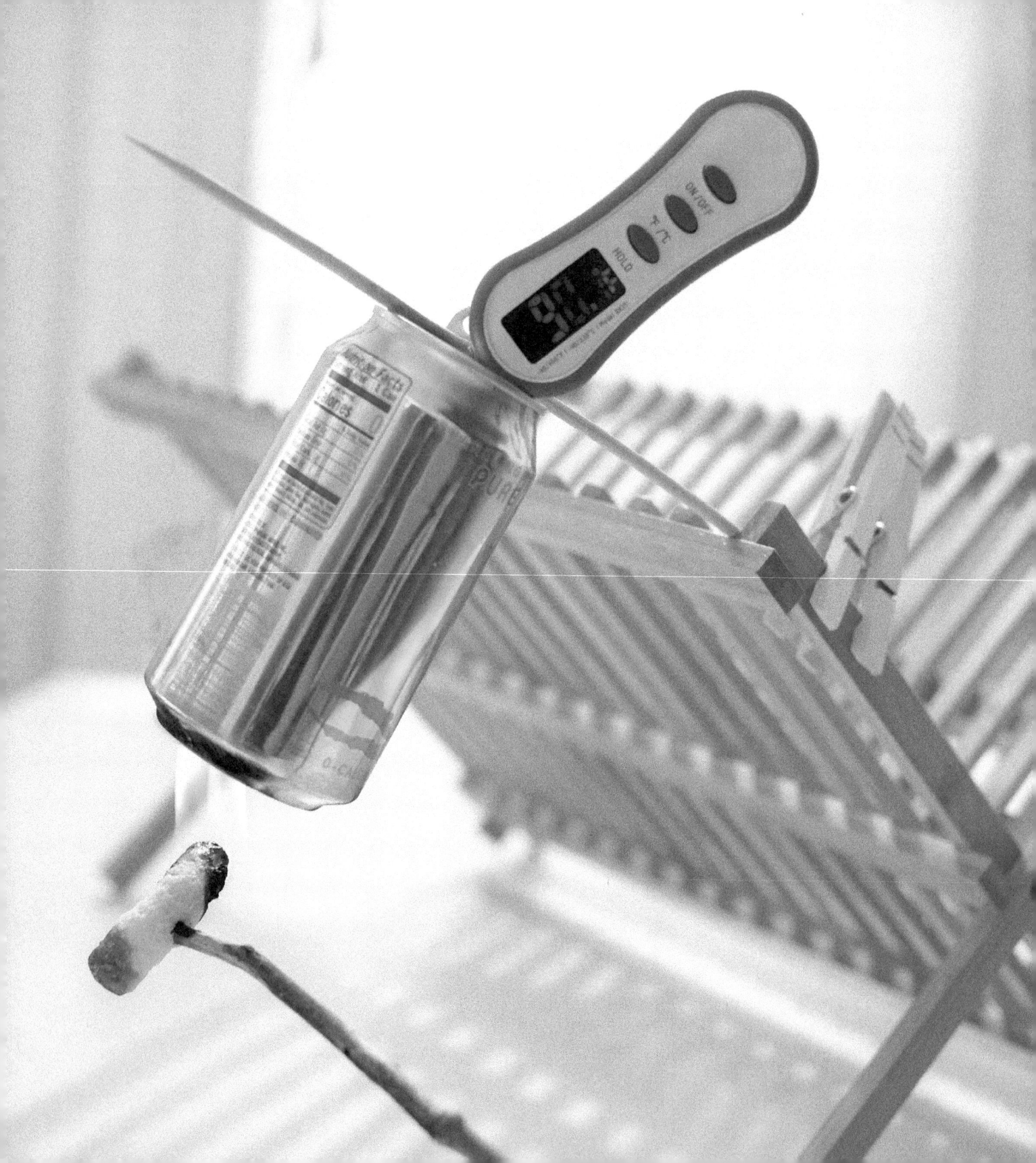

YOUR SNACK IS ON FIRE: FLAMING CHEESE PUFFS

LEVEL OF DIFFICULTY: ADVANCED
MESS-O-METER RATING: MINOR MESS
BEST EATEN FOR: SNACK
PREP TIME: NONE
FULL TIME: 10 MINUTES TO BUILD THE CALORIMETER AND 5 MINUTES TO BURN EACH FOOD
MAKES: SNACKS FOR 1 TO 2 PEOPLE

 A calorie is a unit of energy in food. Food chemists build calorimeters to measure the exact number of calories in certain foods. We're going to build our own calorimeter to answer this burning question: Does a cheese puff or a marshmallow have more calories?

Cautions: Open flame is very dangerous. Wear goggles and have an adult manage the matches and flaming snacks.

THE STEPS

1. Set up your calorimeter. Thread the heat-safe dowel through the hole in the tab of the empty soda can. Use the clips to secure the dowel to the rack so that the empty soda can dangles over the kitchen counter with at least 6 inches beneath it.

2. Pour 7 tablespoons of water into the empty soda can.

TOOLS:

- Heat-safe dowel (for example, a bamboo skewer soaked in water)
- Empty soda can
- 2 to 4 clips, such as clothespins
- Rack (like an empty dish rack)
- Candy thermometer
- Marshmallow roasting stick
- Matches

INGREDIENTS:

- 7 tablespoons water
- 1 snack-sized package of cheese puffs
- 10 marshmallows

CONTINUED

Hypothesis: Which food do you think has more calories, a cheese puff or a marshmallow? Why?

_____

_____

_____

_____

Observations: List the water temperatures before and after each snack fire.

_____

_____

_____

_____

Results: Which snack item heated the water in the can the most?

_____

_____

_____

_____

3. Carefully insert the thermometer into the soda can.

4. Spear one cheese puff on the dowel.

5. Observe the temperature of the water in the can.

6. With an adult's supervision, use the matches to light the cheese puff on fire.

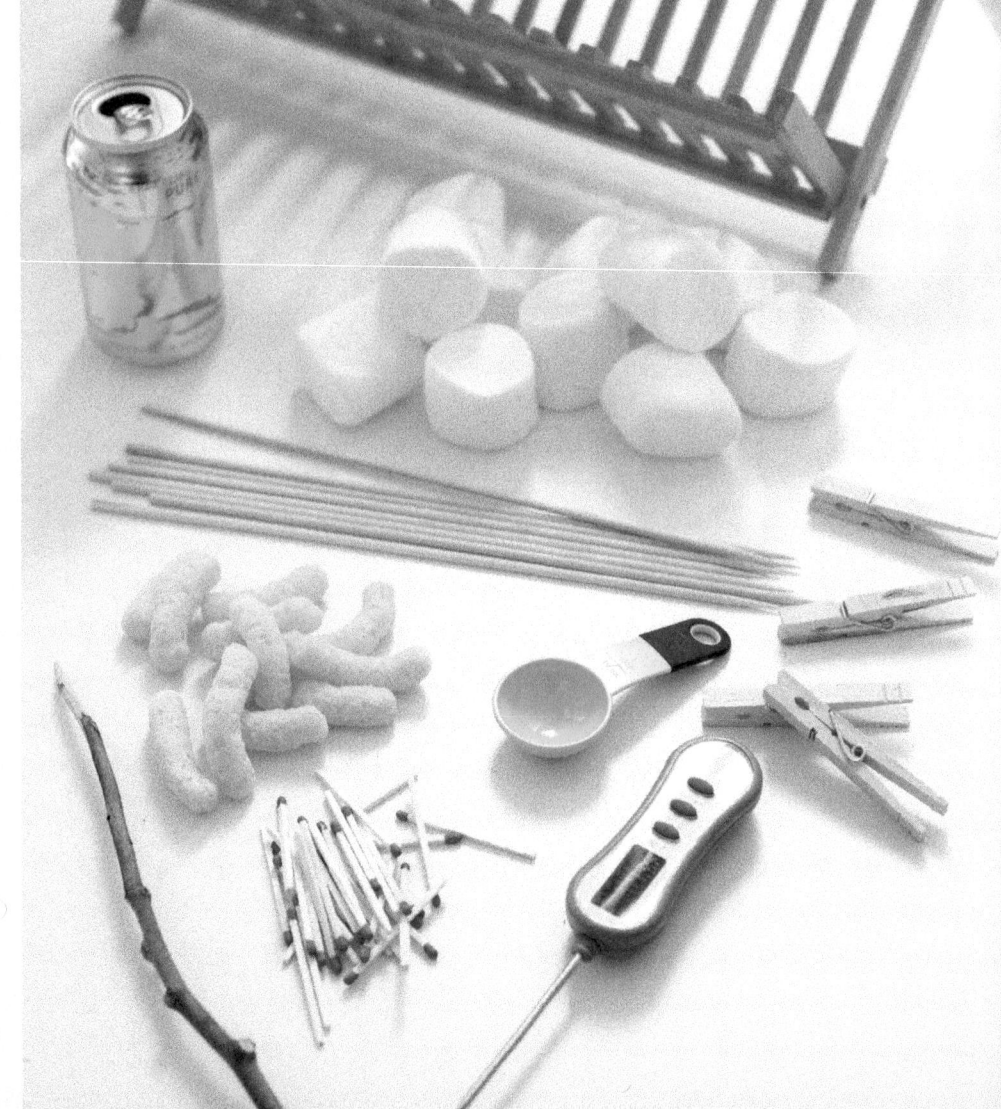

7. Hold the flaming cheese puff directly under the soda can until it is totally burned up, or until there is so little left that it can't stay on the stick. Do not touch the hot cheese puff.

8. Observe the temperature of the water in the can.

9. Repeat steps 2 to 8 with the marshmallow.

10. Record your results.

11. If you're hungry, eat the rest of the cheese puffs and marshmallows. Yum!

The Hows and Whys We eat food because it contains energy to power our bodies. The energy in food comes from the chemical bonds within the molecules of the food. When food burns, the chemical bond energy transforms into heat energy. The heat energy warmed the water in the can. The snack that heated the water the most had the most calories.

STEAM CONNECTION: Whenever a processed food product arrives at a grocery store, it is legally required to have a food label. Food scientists use calorimeter technology to measure exactly how many calories are in foods like cheese puffs, marshmallows, chips, and pretzels.

Turn It Up! Every degree that the water in the can increased represents 0.1 calories. How many calories were in each snack item? Do the math.

HIDDEN PIGMENTS: SPINACH CHROMATOGRAPHY

TOOLS:

- 1 white cone coffee filter
- Scissors
- Ruler
- Pencil
- Coin
- Clear cup
- Tape
- Isopropyl alcohol (rubbing alcohol)
- Paper towel

INGREDIENTS:

- 2 cups spinach leaves
- 1 cup sliced strawberries
- 2 tablespoons of your favorite salad dressing

LEVEL OF DIFFICULTY: ADVANCED
MESS-O-METER RATING: MINOR MESS
BEST EATEN FOR: LUNCH OR DINNER
PREP TIME: NONE
FULL TIME: 30 MINUTES
MAKES: 2 SIDE SALADS

 Pigment molecules give vegetables their vibrant colors. We know that most leaves are green, but do spinach leaves have only green colors? If you separate out the colors in a leaf, how many colors will you see?

Cautions: Isopropyl alcohol can sting if it splashes in your eyes. Wear goggles and have an adult pour the isopropyl alcohol.

THE STEPS

1. Cut a long strip from the coffee filter. The strip needs to be about 1 centimeter longer than the cup and about 2.5 centimeters wide.

2. Use the ruler to measure 2 centimeters up from the bottom of your filter paper strip. At this point, draw a horizontal line across your filter paper strip with the pencil.

3. Lay a spinach leaf flat on top of your filter paper strip, directly over your pencil line. Carefully roll the coin over the spinach, following the pencil line. Lift up the spinach. You should have a thin green line directly on top of your pencil line, 2 centimeters above the bottom of your filter paper strip.

4. Tape the top of your filter paper strip to the pencil. The pencil will rest horizontally across the top of the cup so that the filter paper strip hangs down, barely touching the bottom of the cup.

5. Have an adult pour 1 centimeter of rubbing alcohol into the bottom of the cup. Do not let the rubbing alcohol touch your filter paper strip anywhere except the very bottom, which should be resting in the rubbing alcohol. The green line from the spinach needs to be above the rubbing alcohol.

6. The rubbing alcohol will travel up the filter paper strip. As it moves through the spinach line, watch what happens to the line. When the liquid is almost to the top of the filter paper strip, take it out of the cup and set it on a paper towel. Record your observations.

Hypothesis: List the colors that you think you will find in a spinach leaf.

_____

_____

_____

_____

_____

Observations: What interesting shapes or lines did you see?

_____

_____

_____

_____

Results: List the colors that you discovered through chromatography.

_____

_____

_____

_____

_____

CONTINUED

The Hows and Whys Spinach leaves have four color molecules: bright green, dark green, orange, and yellow. The orange and yellow colors are lighter so the dark green colors cover them up. Even though we normally can't see these hidden colors, they have an important job. The pigments help leaves capture sunlight energy and turn it into sugar in a process called **photosynthesis**.

STEAM CONNECTION: Chromatography is a tool used in chemistry labs to identify important molecules. Vitamins are important food molecules that help the body's **enzymes** do their jobs. You'll learn more about enzymes in the next chapter.

Turn It Up! Use the remaining spinach leaves to make a delicious strawberry-spinach salad. Place half of the leaves on each of the 2 plates. Place half of the strawberries on top of each of the 2 plates. Drizzle half of the dressing on top of each salad. Enjoy the flavors of those pigments.

EXTRA FEATURE: WOK & ROLL WITH CHEF HERVÉ THIS

Hervé This ("Tees") is a world-famous French chef. He's also a chemist. In 1988, This cofounded a new field of science called molecular gastronomy that focuses on food molecules and the changes that happen to them during the cooking process. He works to find and separate out the specific molecules that give foods their flavors. Then he uses what he's learned to figure out how to make **artificial** flavors in his laboratory.

He is also known for discovering powders and foams that offer the food energy we need, including **starches** and proteins, like regular food does. A dinner made from powders, foams, and flavor droplets might sound boring, but chefs have used This's discoveries to do pretty awesome things in the kitchen.

Today, Hervé This works on Note by Note (NbN) cuisine, a company that uses special flavor molecules with nutrients to create familiar-tasting dishes with very little waste. This believes that NbN is the solution to world hunger—molecules are smaller, lighter, and less likely to spoil than natural foods. It may be 20 years or more before we see these technologies in household kitchens. For now, it's hard to beat the all-natural crunch of a fresh, sweet apple, even in the world's fanciest kitchens.

Chapter Five

ENGINEERING

Welcome to the wild world of engineering! Engineers ask questions to understand the world's problems, and they use mathematics to design solutions. Engineers often work in groups—sharing ideas, showing designs, and arguing about possible explanations and solutions.

Chemical engineering is important in the kitchen. Many of the foods that we enjoy are made using engineering discoveries. For example, the many different flavors of cheese in the grocery store came from chemical engineering discoveries. Food scientists experiment to discover how to keep food fresh, tasty, and appealing. In this chapter, you will experiment with chemical engineering in fruits, vegetables, cheese, French toast, pancakes, and cookies. You will also use mechanical engineering to build a solar oven and three different edible structures—one of which is a simple machine.

Because the world is full of tricky problems that aren't easy to solve, engineers struggle and fail a lot. Mistakes and unexpected results are part of engineering. Remember that if you have to do an engineering project more than once, you're probably doing it right.

PINEAPPLE PROBLEMS: PROTECTING YOUR PROTEINS

TOOLS:

- 1 (4-cup) liquid measuring cup, clear and heat-proof
- Microwave oven
- Measuring cups
- Spoon
- 4 clear 8-ounce cups
- Knife
- Paper towels

INGREDIENTS:

- 4 cups water, divided
- 1 (3-ounce) package Jell-O
- ¼ cup dried pineapple
- ¼ cup canned pineapple
- ¼ cup fresh pineapple

LEVEL OF DIFFICULTY: MEDIUM
MESS-O-METER RATING: MINOR MESS
BEST EATEN FOR: SNACK OR DESSERT
PREP TIME: NONE
FULL TIME: 10 MINUTES TO MAKE JELL-O, PLUS 30 MINUTES TO MAKE OBSERVATIONS
MAKES: 4 SERVINGS OF FRUITY JELL-O

 Did you know that cooks never put fresh pineapple in old-fashioned Jell-O salad? If they did, the whole dessert would melt into a puddle! Now let's find out: Will canning or drying pineapple make this fruit work for a Jell-O chef?

Cautions: Adult supervision is needed when you are handling hot liquids and sharp utensils.

THE STEPS

1. Measure 2 cups of water in the liquid measuring cup. Microwave for 2 minutes.

2. Add 1 package of Jell-O to the 2 cups of hot water in the liquid measuring cup and stir until the Jell-O dissolves.

3. Add 2 cups of cold water to the Jell-O solution.

4. Pour the Jell-O into the 4 cups.

5. Cool the Jell-O in the refrigerator for about an hour, until it is almost set.

6. While the Jell-O is cooling, dice each type of pineapple separately. Place the canned and fresh pineapple on a paper towel so that it is well drained.

7. When the Jell-O has almost set, place ¼ cup of diced dried pineapple flat on the gelatin in one cup.

8. Repeat with the canned and fresh pineapple. Be sure that the canned pineapple is well drained and that no extra water touches the Jell-O.

CONTINUED

Hypothesis: Predict which type of pineapple (fresh, dried, or canned) will give your Jell-O the fewest problems.

_____
_____
_____
_____
_____

Observations: What happened with your pineapple at 5, 15, and 30 minutes? Be sure to look at your experiment from the side and to note any watery liquid that appears on the surface of the Jell-O.

_____
_____
_____
_____
_____

Results: Which type of pineapple worked best?

_____
_____
_____
_____
_____

9. You now have 1 cup of Jell-O and dried pineapple, 1 cup of Jell-O and canned pineapple, and 1 cup of Jell-O and fresh pineapple. The fourth cup of Jell-O is the control.

10. Study all 4 cups after 5-, 10-, and 30-minute intervals. What is happening to the gelatin under each slice of pineapple? Record your observations and results.

The Hows and Whys Pineapple contains a gelatin-destroying molecule called protease. When pineapple is dried or canned, it gets hot. The heat denatures, or destroys, protease and it stops working. Any time pineapple is heated above 108°F, its proteins will denature and it will no longer present a problem for Jell-O chefs.

STEAM CONNECTION: Scientists working in the field of chemistry are often challenged to limit the impact of problem molecules. In this lab, you searched for a solution to the problems caused by protease molecules in pineapple.

Turn It Up! Jell-O salads prepared in shaped pans called *molds* were popular in the 1960s and 1970s. Search online for "Jell-O mold recipes with fruit" and select the recipe that looks best to you. Prepare the recipe for a vintage dessert to impress your family and friends.

SNEAKY ZUCCHINI: IMITATION APPLE PIE

LEVEL OF DIFFICULTY: ADVANCED
MESS-O-METER RATING: MINOR MESS
BEST EATEN FOR: DESSERT
PREP TIME: 10 MINUTES TO SLICE THE ZUCCHINIS AND APPLES
FULL TIME: 1 HOUR TO MAKE THE PIES, 30 MINUTES TO BAKE THE PIES, 1 HOUR FOR THE PIES TO COOL, AND 10 MINUTES FOR TASTE TESTS
MAKES: 2 PIES

 Food scientists engineer imitation foods that trick our senses. In this lab, you will be creating two versions of apple pie and running taste tests. Can people tell the difference between real and imitation apple pie? If so, why?

Cautions: Adult supervision is needed when using the oven, stove, and knife.

TOOLS:

- Knife for slicing apples and zucchini
- Spoon
- 1 large frying pan
- Spatula
- Stove
- 2 large bowls
- Measuring cups and spoons
- Food processor
- 2 cookie sheets, rimmed
- Oven
- Oven mitts

THE STEPS

1. Peel the 2 or 3 large zucchini and cut them in half the long way. Using the spoon, scoop out the seeds in the middle of the zucchini. Slice the zucchini the short way, making about 6 cups of ¼-inch thick zucchini slices.

2. Put the zucchini slices, 3 tablespoons of lemon juice, and a pinch of salt in the frying pan.

CONTINUED

INGREDIENTS:

FOR THE PIES

- 2 or 3 large zucchini
- Juice from 1 lemon (about 6 tablespoons)
- 2 pinches salt, divided
- 4 to 7 large apples
- 2½ cups brown sugar, divided
- 3 teaspoons ground cinnamon, divided
- 2 pinches ground nutmeg, divided
- 4 teaspoons cream of tartar, divided
- ½ cup all-purpose flour, divided
- 2 prepared pie shells in tins

FOR THE TOPPING

- 2 cups flour
- 1 cup sugar
- ½ cup brown sugar
- 3 teaspoons cinnamon
- 1 teaspoon salt
- 12 tablespoons unsalted butter, cut into ½-inch cubes and kept cold in the refrigerator

3. Cook the zucchini in the frying pan over medium heat. Do not brown the zucchini slices, but cook them until they are soft. Use the spatula to stir them often.

4. Turn off the heat.

5. While the zucchini slices cool, slice 4 to 7 apples so that you have about 5 cups of ¼-inch thick apple slices.

6. Put the zucchini slices in one large bowl and the apple slices in another large bowl.

7. Add 3 tablespoons of lemon juice and a pinch of salt to the bowl with apple slices.

8. Add 1¼ cups of brown sugar to each bowl.

9. Add 1½ teaspoons of ground cinnamon to each bowl.

10. Add a pinch of ground nutmeg to each bowl.

11. Add 2 teaspoons of cream of tartar to each bowl.

12. Add ¼ cup of flour to each bowl.

13. Stir the mixtures until both the zucchini pie filling and the apple pie filling are combined well.

14. Pour the imitation (zucchini) pie filling into one pie crust.

15. Pour the apple pie filling into the other pie crust.

16. Preheat the oven to 400°F.

17. In the food processor, pour in 2 cups of flour, 1 cup of sugar, ½ cup of brown sugar, 3 teaspoons of cinnamon, and 1 teaspoon of salt and pulse twice to combine.

18. Add the butter to the mixture. Pulse 5 to 12 times, until the mixture looks like damp sand.

19. Pour half of this mixture over each of the two pies.

20. Place the pies on the rimmed cookie sheets.

21. Bake the pies for 30 minutes in the 400°F oven.

22. Set the pies out of the oven to cool for 1 hour.

23. Invite friends and family to taste the two pies. Do not tell your tasters which pie is which, but keep track so that you know! Taste the pies yourself. Record your observations and results.

The Hows and Whys As food scientists, we can use the power of spices to trick our senses. When we taste cinnamon, nutmeg, lemon, and sugar, we think of apple pie—even if there aren't any apples in our dessert. Some imitation foods are less healthy than the real thing, but it doesn't have to be that way. You used a healthy vegetable! Were your taste-testers able to tell the difference?

STEAM CONNECTION: When real ingredients aren't practical or affordable for a recipe, food chemists create tasty imitations. Understanding the science of flavor helps engineers design convincing imitations.

Turn It Up! The next time you have spaghetti or sloppy joes for dinner, add 1 cup of pureed squash to the tomato sauce. This is a great way to sneak an extra vegetable into your day.

Hypothesis: Predict whether people will be able to tell the difference between real and imitation apple pie.

_____
_____
_____
_____

Observations: How did the imitation pie compare to the real apple pie?

_____
_____
_____
_____

Results: Could people tell the difference between the two pies?

_____
_____
_____
_____
_____

THAT'S SO CHEESY: PLAYING WITH WHEY

TOOLS:

- Colander
- Bowl
- Cheesecloth
- Candy thermometer with clip
- Pot
- Stove
- Spoon
- Measuring cup and spoons
- Slotted spoon
- Plate
- Silicone candy mold
- 4-cup liquid measuring cup
- Plastic storage container

INGREDIENTS:

- ½ gallon whole milk (pasteurized is okay, but not ultra-pasteurized)
- ⅓ cup white vinegar
- 1 teaspoon salt

LEVEL OF DIFFICULTY: MEDIUM
MESS-O-METER RATING: MEDIUM MESS
BEST EATEN FOR: LUNCH, DINNER, OR SNACK
PREP TIME: 5 MINUTES TO GATHER SUPPLIES
FULL TIME: 1 HOUR TO MAKE THE CHEESE, PLUS 1½ HOURS FOR CHEESE TO SET
MAKES: 1 SMALL BLOCK OF CHEESE (ABOUT ¾ CUP)

Food scientists make cheese by taking the water out of milk. In this experiment, you will use heat and acid to make a basic cheese called **queso fresco**. *Queso fresco* is delicious fried, grilled, or spread on bread, but it won't melt. What gives cheese its texture and flavor?

Cautions: Adult supervision is needed while using the stove and handling the hot milk.

THE STEPS

1. Put the colander in the bowl.

2. Put 4 layers of cheesecloth on top of the colander.

3. Clip your thermometer to the side of the pot so that it goes inside the pot without touching the bottom or sides.

4. Pour ½ gallon of whole milk into the pot.

5. Heat the milk over medium heat until it is between 165°F and 185°F, then turn off the stove. This will take

about 15 minutes. Gently stir the milk at least once every minute.

6. Slowly add ⅓ cup of white vinegar to the milk in 5 or 6 small splashes, gently stirring the milk after each splash.

7. The liquid **whey** in the pot will turn clear, and you will see white cheese **curds** will start to form. Record your observations. Let the curds and whey sit for 15 minutes to finish separating.

8. Use the slotted spoon to scoop the curds out of the whey. Set each spoonful of curds onto the cheesecloth in the colander. Sprinkle the top of the cheese with the salt.

9. Leave the curds in the colander for 20 minutes to let the whey drain away. Cover the top of the colander with a plate to keep the cheese clean.

10. Use a spoon to scoop out a small portion of cheese and press it into the silicone candy mold. At this stage, your cheese will take any shape, so your cheese will match the shape of the mold.

11. Pull the corners of the cheesecloth together and tie them in a knot. The remaining cheese will be a ball inside the cheesecloth. Keep the cheese in the colander.

12. Place the 4-cup liquid measuring cup on top of the cheese to press out more of the whey. Leave the cheese to set for 1½ hours.

13. Open the cheesecloth and, with clean hands, move the cheese to the plastic storage container. Taste the cheese and record your results. Store your cheese in the refrigerator.

Hypothesis: What texture and flavor will homemade cheese have?

_____
_____
_____
_____
_____

Observations: Record what you see and smell as you make the cheese.

_____
_____
_____
_____
_____

Results: Describe the texture of your cheese. What did it taste like?

_____
_____
_____
_____
_____

CONTINUED

The Hows and Whys When you heated the milk and added acidic vinegar, the milk separated into curds (made of protein and fat) and whey (water). Pressing the cheese squeezed out the extra whey and gave your cheese its shape. These steps give cheese its texture and flavor.

STEAM CONNECTION: Cheese is one of the first food engineering discoveries. Historians believe that cheese was discovered accidentally when someone accidentally let their milk go sour—gross! Today, modern technologies have made it possible to make many flavors and textures of cheese.

Turn It Up! If you enjoyed making cheese, consider purchasing a cheese-making kit. Kits for making popular cheeses, including mozzarella and cheddar, are easy to purchase online.

THE TALLEST TASTE:
MAILLARD REACTION FRENCH TOAST

LEVEL OF DIFFICULTY: MEDIUM
MESS-O-METER RATING: MINOR MESS
BEST EATEN FOR: BREAKFAST
PREP TIME: 5 MINUTES TO MIX THE FRENCH TOAST BATTER
FULL TIME: 20 MINUTES
MAKES: BREAKFAST FOR 2 PEOPLE

 Did you ever wonder how French toast gets that beautiful golden-brown color (and rich, savory flavor) when you cook it? In this lab, you'll be exploring the Maillard reaction, an important chemical change that happens when proteins and sugars are heated together. How hot does French toast need to be to start to brown?

Cautions: Adult supervision is required for cooking on the stove.

THE STEPS

1. To make the French toast batter, use the whisk to mix the 2 eggs, ½ cup of milk, ½ teaspoon of vanilla, and ¼ teaspoon of nutmeg in the shallow bowl.

2. Put 1 tablespoon of butter in each of the 2 frying pans.

3. Turn one stove burner to low.

CONTINUED

TOOLS:

- Wide, shallow bowl
- Measuring cups and spoons
- Whisk or fork
- 2 frying pans
- Stove
- Spatula

INGREDIENTS:

- 2 eggs
- ½ cup milk
- ½ teaspoon vanilla
- ¼ teaspoon nutmeg
- 2 tablespoons butter
- 4 slices bread

Hypothesis: How hot do you think the pan needs to get for the Maillard reaction to begin?

_____

_____

_____

_____

Observations: Note what you see in each of the two frying pans.

_____

_____

_____

_____

_____

Results: Which setting on the stove gave you the most beautifully golden-brown French toast?

_____

_____

_____

_____

_____

4. Turn the second burner to medium.

5. Dip one piece of bread in the French toast batter. Check to see that it is completely coated. Place the piece of bread on one side in the frying pan on low. Repeat with a second piece of bread so that there are two pieces of bread in the frying pan.

6. Repeat step 5 for the frying pan on medium.

7. Watch the four slices of French toast carefully, flipping them over if and when they turn golden brown on the bottom.

8. Record your observations and results.

> **The Hows and Whys** The Maillard reaction begins in earnest at 350°F. Frying, grilling, roasting, and baking foods above 350°F begins a chain reaction in which sugars chemically react with proteins to produce molecules that change the food's color and flavor.

STEAM CONNECTION: Thermometers are a technology that help scientists discover the best temperatures for chemical reactions.

Turn It Up! Experiment with cooking French toast in a frying pan over a burner turned to the high setting. How hot can you get the French toast before burning it?

HAPPY APPLES:
CINNAMON APPLE PANCAKES

LEVEL OF DIFFICULTY: EASY
MESS-O-METER RATING: MINOR MESS
BEST EATEN FOR: BREAKFAST
PREP TIME: 5 MINUTES TO SLICE THE APPLE AND BOIL WATER
FULL TIME: 15 MINUTES
MAKES: 1 APPLE SNACK OR PANCAKES FOR 4

 Most people don't like the look or taste of brown apples. The engineering challenge in this experiment is to discover the best technique for slowing down the process that makes apples brown when they are exposed to air. Does lemon juice or hot water slow down apple browning better?

Cautions: Adult supervision is required when slicing apples, boiling water, and preparing pancakes on the stove.

THE STEPS

1. Pour 2 cups of water into a teakettle and have an adult turn on the stove to boil the water.

2. Slice an apple into 4 quarters.

3. Slice each apple quarter into 2 to 8 thin slices.

TOOLS:

- Measuring cups
- Teakettle or pot for boiling water
- Cutting board
- Knife
- 3 bowls

FOR PANCAKES (OPTIONAL)

- 4-cup liquid measuring cup
- Measuring spoons
- Spoon
- Pancake skillet or griddle
- Stove
- Spatula

CONTINUED

INGREDIENTS:

- 2 cups water
- 1 apple
- Juice from lemon (about ½ cup juice)

FOR PANCAKES (OPTIONAL)

- Pancake mix
- 1 teaspoon cinnamon
- Water, milk, and/or eggs

Hypothesis: Which will slow apple browning better, lemon juice or boiling water?

_____

_____

_____

_____

_____

_____

_____

4. Place ¼ of the apple slices in 1 of the 3 bowls. Cover with boiling water.

5. Place ¼ of the apple slices in the second bowl. Cover with lemon juice.

6. Place ¼ of the apple slices in the third bowl. Do not add anything else to the bowl. This is your control.

7. Let the apple slices sit in the bowls for 5 minutes. Eat 1 of the leftover apple slices while you wait.

8. After 5 minutes, record your observations, noting how the apples look and taste. If no browning has occurred in the control bowl, wait another 5 minutes before recording your results.

The Hows and Whys Apples, like many fruits and vegetables, are full of enzymes. Enzymes are proteins that speed up chemical reactions. When a sliced apple turns brown, an enzyme has **catalyzed** a reaction between the apple's natural chemicals and oxygen in the air. The acid in lemon juice and the heat in boiling water both denature the enzyme and slow down the browning reaction.

STEAM CONNECTION: When food looks unappetizing, that's a big problem. Technologies that stop unwanted enzymatic browning help food scientists make food look as delicious as it tastes.

Turn It Up! Use the leftover apple slices to make apple cinnamon pancakes. Following the pancake mix directions, prepare one batch of pancake batter. Mix 1 teaspoon of cinnamon into the batter. With adult supervision, cook the pancakes, placing one thin apple slice on top of each pancake once bubbles have appeared in the batter (but before you flip it over).

Observations: What do you notice happening to the apple slices?

_____

_____

_____

_____

_____

Results: Which apples were the most brown? The least brown? How did the apples taste?

_____

_____

_____

_____

_____

EGG FOAM SCIENCE: MAGNIFICENT MACAROONS

LEVEL OF DIFFICULTY: MEDIUM
MESS-O-METER RATING: MINOR MESS
BEST EATEN FOR: DESSERT
PREP TIME: NONE
FULL TIME: 25 MINUTES TO PREPARE AND 12 MINUTES TO BAKE
MAKES: 2 OR 3 DOZEN COOKIES

Chefs often whisk egg whites into foam. This delivers delicious textures into baked foods. In this lab, you'll build two egg foams and compare their strength and stability. You'll also see the Maillard reaction again—this time, in the oven, as your cookies turn a beautiful golden brown. Will adding sugar before or after whisking egg whites make the strongest foam?

 **Cautions:** Adult supervision is required when using the electric mixer and the oven. Do not eat uncooked eggs.

THE STEPS

1. Preheat oven to 375°F and place a sheet of parchment paper on each of the two cookie sheets.

2. In the mixing bowl, beat 3 egg whites with the electric mixer on high speed until they form shiny, stiff peaks (6 or 7 minutes).

CONTINUED

TOOLS:

- Oven
- Oven mitts
- 2 cookie sheets
- Parchment paper
- Large mixing bowl
- Electric mixer (handheld or stand mixer)
- Measuring cups
- Rubber spatula
- Spoon
- Timer

INGREDIENTS:

- 6 egg whites, divided
- 2 pinches salt, divided
- 1 cup sugar, divided
- 2 cups unsweetened coconut (shredded or flaked)

Hypothesis: Write your prediction about how to make the strongest egg foam: by adding sugar before or after whisking eggs.

_____

_____

_____

_____

Observations: Record the heights of your two egg foam towers here.

_____

_____

_____

_____

Results: Which egg whisking technique made the stronger foam?

_____

_____

_____

_____

EGG FOAM SCIENCE: MAGNIFICENT MACAROONS, CONTINUED

3. Gently stir in 1 pinch of salt and ½ cup of sugar.

4. Pour this mixture onto a clean plate, using the rubber spatula to make the pile of egg foam as tall as possible.

5. Put the remaining 3 egg whites and pinch of salt in the same mixing bowl and begin mixing on high speed.

6. As soon as the egg whites start to foam, slowly sprinkle the remaining ½ cup of sugar while the mixer is still running.

7. Continue to mix on high speed until the eggs form shiny, stiff peaks (9 or 10 minutes total).

8. Pour this mixture onto a clean plate, using the rubber spatula to make the pile of egg foam as tall as possible.

9. Use a ruler to measure both towers of foam and record your observations.

10. Take your tallest foam tower and return it to the mixing bowl.

11. Add 2 cups unsweetened coconut to the mixing bowl and stir it in gently.

12. Drop the egg foam onto the parchment-lined cookie sheets by the spoonful. You can make any size cookie you like.

13. Bake the cookies for 12 minutes. Turn off the oven and let the cookies cool to room temperature.

The Hows and Whys Egg whites are made of proteins that form foam when whisked. Sugar molecules can stabilize the foam if they are present when it's forming but not if the sugar is added after the foam forms. Adding sugar before beating eggs makes the strongest foam.

STEAM CONNECTION: Understanding what makes proteins denature—as the egg proteins did when they were whisked—helps biologists and doctors who study proteins in nature and in the human body.

Turn It Up! If you're confident with your egg foam skills, look in the Resources to find a recipe for chocolate raspberry soufflé. The pectin in the raspberry jam helps keep the egg foam strong and increases your chances of soufflé success.

A BRIGHT FUTURE: SOLAR-POWERED S'MORES

TOOLS:

- 2 cardboard boxes with hinging tops, like pizza boxes
- Scissors or a utility knife
- 2 pieces of black construction paper
- Aluminum foil
- Clear plastic wrap
- Tape
- Weather thermometer

INGREDIENTS:

- Graham crackers
- Chocolate squares
- Marshmallows

Hypothesis: Predict whether black paper or aluminum foil lining will make a hotter solar oven.

_____

_____

_____

_____

_____

LEVEL OF DIFFICULTY: EASY
MESS-O-METER RATING: MINOR MESS
BEST EATEN FOR: SNACK OR DESSERT
PREP TIME: 3 MINUTES TO CUT A WINDOW INTO THE TOP OF A BOX
FULL TIME: 1 OR 2 HOURS, DEPENDING ON THE TEMPERATURE OUTSIDE
MAKES: AS MANY S'MORES AS YOU CAN EAT

 Every day, Earth soaks up light and heat, or **radiation**, from the Sun. Most of this energy is not used, but engineering can solve this problem by building a solar oven to bake s'mores. You will build a solar oven out of a cardboard box and line the box with materials that will help hold the Sun's heat. What colors maximize the heat inside your oven?

Cautions: Step 1 needs to be done by an adult.

THE STEPS

1. Have an adult draw and cut out a window on the top of each cardboard box, leaving 1 to 2 inches of a frame around the window.

2. Line the inside of one box with black construction paper.

3. Line the other box with aluminum foil.

4. Cover the windows with clear plastic wrap and tape the plastic wrap to the box.

5. In each box, set ½ of a graham cracker flat on the bottom.

6. Place a chocolate square on top of each graham cracker.

7. Place a marshmallow on top of each chocolate square.

8. Close the lids on both solar ovens.

9. Put both solar ovens outside in a hot sunbeam.

10. Use the thermometer to measure the heat in each oven every 5 minutes for 1 hour. Record your observations and results.

The Hows and Whys Black objects **absorb** radiation. Reflective objects, including aluminum foil, reflect radiation. A solar oven with a black bottom liner and reflective top liner should get the hottest.

STEAM CONNECTION: Earth gets a lot of solar radiation every day. Solar panels—flat black devices that make electricity—are a technology created to capture the energy in solar radiation for human use.

Turn It Up! Now that you've tested a basic solar oven, engineer a design that uses the light-absorbing power of black paper and the reflective power of aluminum foil. Play with your s'mores recipes. You can substitute any kind of cookie for the graham crackers and any flavor of chocolate (including peanut butter chocolate cups) for the plain chocolate squares.

Observations: Record what you see while the s'mores are baking in the oven. What melts first?

_____

_____

_____

_____

_____

_____

_____

_____

Results: Did the black lining or the aluminum foil lining make the hotter solar oven? What was the temperature inside each oven?

_____

_____

_____

_____

_____

_____

_____

_____

_____

_____

DUELING DOMES: GEODESIC GUMDROP DESIGNS

LEVEL OF DIFFICULTY: EASY

MESS-O-METER RATING: MEDIUM MESS

BEST EATEN FOR: DESSERT

PREP TIME: NONE

FULL TIME: 30 MINUTES ONE DAY, THEN 15 MINUTES THE NEXT DAY

MAKES: 2 DESSERTS

Architectural engineers have to think about how the parts of a building will carry stress. The stress from heavy buildings pulls on the beams holding up the buildings. Do you think that squares or triangles will be best for distributing stress so that a building can stay standing over time?

Cautions: Always push sharp objects, including toothpicks, away from yourself. If you want to eat any gumdrops, wash your hands before beginning the experiment. Do not eat gumdrops that have had toothpicks in them, in case there are splinters that were left behind.

THE STEPS

1. Place 4 gumdrops in a square on a clean plate.

2. Press 4 toothpicks firmly into the 4 gumdrops to connect them in a square shape. The gumdrop square should lay flat on the plate.

CONTINUED

TOOLS:

- 2 plates
- 1 box of toothpicks

INGREDIENTS:

- 1 large bag of gumdrops (or another gummy candy, such as Dots®, fruit snacks, or Swedish Fish®)

Hypothesis: Will a dome design with squares or triangles be stronger?

_____

_____

_____

_____

_____

_____

_____

Observations: How many books could each dome hold?

_____

_____

_____

_____

_____

Results: Which dome design was stronger?

_____

_____

_____

_____

_____

_____

_____

3. Place 1 toothpick in each gumdrop so that it sticks straight up.

4. Place 4 more gumdrops on top of the toothpicks that are sticking up, then use 4 additional toothpicks to connect the new gumdrops in a square. You should have a gumdrop cube.

5. Set aside your gumdrop cube to set overnight.

6. Place 5 gumdrops in a pentagon on a second clean plate.

7. Press 5 toothpicks firmly into the 5 gumdrops to connect them in a pentagon shape. The gumdrop pentagon should lay flat on the plate.

8. Press 2 toothpicks into 2 neighboring gumdrops so that they come together at the top to form a triangle. Press 1 gumdrop firmly onto both toothpick tops to finish the triangle.

9. Repeat until you have 5 new gumdrops suspended over 5 toothpick triangles.

10. Use 5 more toothpicks to firmly connect the 5 new gumdrops in a second pentagon.

11. Press 1 toothpick into each of the 5 new gumdrops so that it is angled toward the center of the pentagon. Firmly insert all 5 toothpicks into a final gumdrop at the center of the top pentagon.

12. Set your dome aside to set overnight.

13. Note: For stress tests tomorrow, you might want to make 3 domes.

14. Once both of your structures have dried so that the toothpicks are firmly stuck inside the gumdrops, test their strength by placing books on top. Which structure can bear more stress? Record your observations and results.

The Hows and Whys From a distance, the dome you made might look like a sphere, but mathematicians know that it is actually a **polyhedron**. In a polyhedron, many flat shapes—in this case, triangles—come together to make a 3-dimensional object, like a pyramid. Polyhedrons are strong because their many flat shapes spread the stress evenly throughout the structure.

STEAM CONNECTION: Engineers often apply mathematics in their designs. In this lab, you used **geometry**, a kind of math focused on shapes.

Turn It Up! Try building a larger dome with more gumdrops. What is the largest base you can make before the dome caves in? Consider an octagon (8 sides), decagon (10 sides), or dodecagon (12 sides).

LAUNCH YOUR LUNCH:
DESIGN A MARSHMALLOW CATAPULT

LEVEL OF DIFFICULTY: EASY
MESS-O-METER RATING: MEDIUM MESS
BEST EATEN FOR: SNACK OR DESSERT
PREP TIME: NONE
FULL TIME: 30 TO 60 MINUTES
MAKES: 1-PLUS CUPS OF HOT CHOCOLATE
WITH MARSHMALLOWS

Engineers build machines to do jobs. One example of a simple machine is a **lever**—a stick resting on a support with a person pushing on one end and an object to move on the other end. When the person pushes one end of the lever, the object attached to the other end moves. What lever design can you engineer that will move a marshmallow through the air and into a cup of hot chocolate?

Cautions: Be mindful of the sharp ends of the skewers. Do not eat any marshmallows that have been skewered; they might have wooden splinters inside.

TOOLS:

- 1 sturdy mug for each mug of hot chocolate that you want to make
- Large table or countertop
- Plastic spoons and forks (at least 2 different shapes)
- Wooden skewers (1 package of at least 12)
- Tape
- 5 rubber bands

INGREDIENTS:

- 1 package large marshmallows
- Hot chocolate mix (enough for each mug of hot chocolate that you want to make)

CONTINUED

Hypothesis: Draw your prediction for the lever design that will move your marshmallow.

Observations: Record what worked and what didn't work.

_____
_____
_____
_____
_____

Results: Draw your final lever design.

THE STEPS

1. Set up the sturdy mug on a large table.

2. Standing at one end of the table, place one marshmallow on a plastic spoon or fork.

3. Holding the end of the handle with one hand, use the other hand to pull the top of the spoon or fork back, then let go suddenly. The marshmallow should fly through the air.

4. Pick up the marshmallow and keep launching it, trying out every plastic spoon or fork. Try to get the marshmallow into the cup.

5. Tape your favorite launching spoon or fork to a skewer using the tape. You should have tape wrapped around the handle of the spoon and the skewer.

6. Think about what kind of catapult machine you could build using marshmallows, skewers, and rubber bands. The machine will hold the bottom end of the spoon-skewer tightly while letting the launch end move back and forth.

7. Start building. Test as you go. You will need to rebuild several times. If you have marshmallows all over your table, you're doing it right! Record your observations as you work.

8. When you are satisfied with your catapult, have an adult prepare a mug of hot chocolate. Launch a marshmallow into the cup and enjoy your treat while you record your results.

The Hows and Whys When you push down on one end of a lever, the other end lifts up. When you pulled back on the spoon holding the marshmallow, this was the same as pushing down on one end of a lever. The marshmallow at the very tip of the spoon moved as a result of your work.

STEAM CONNECTION: Physics is the field of science that explores matter (like marshmallows and levers) and energy (like the work you put into bending back the spoon to launch marshmallows). When engineers apply physics in their designs, they use mathematics to predict how their machines will work.

Turn It Up! Try launching other foods. How does your lever work with something heavier than a marshmallow, like a grape? Can you launch a mini marshmallow farther than a regular marshmallow?

HOME SWEET HOME: BUILD A GINGERBREAD HOUSE

TOOLS:

- 2 large mixing bowls
- Measuring cups and spoons
- Electric mixer
- Rubber spatula
- Plastic wrap
- Card stock
- Tape
- Parchment paper
- 2 cookie sheets
- Cutting board
- Rolling pin
- Oven
- Oven mitts

LEVEL OF DIFFICULTY: EASY (IF YOU USE PREMADE GINGERBREAD AND THE TEMPLATE PROVIDED) OR ADVANCED (IF YOU MAKE YOUR OWN ICING, BAKE YOUR OWN GINGERBREAD, AND ENGINEER YOUR OWN HOUSE DESIGN)

MESS-O-METER RATING: MEDIUM MESS

BEST EATEN FOR: DESSERT

PREP TIME: FOR YOUNGER SCIENTISTS, PARENTS SHOULD PREPARE THE GINGERBREAD DOUGH AND CUT OUT THE HOUSE TEMPLATE ON CARD STOCK THE NIGHT BEFORE BUILDING. OLDER SCIENTISTS WILL ENJOY MAKING THEIR OWN DOUGH, RESEARCHING ARCHITECTURAL STYLES, AND DESIGNING A UNIQUE GINGERBREAD HOUSE.

FULL TIME: 3 HOURS

MAKES: 1 HOUSE (4 TO 8 SERVINGS)

Architecture is an artistic field of engineering. It involves designing and building homes and other structures. In this lab, you will use your knowledge of shapes and stability to design and build a gingerbread house. What shapes will make the most stable roof and walls for a gingerbread house?

 **Cautions:** Adult supervision is required when using the oven.

THE STEPS

1. Make the gingerbread dough and chill it overnight (recipe on page 110).

2. Research house shapes by walking around your neighborhood, looking through books on architecture, or doing an online search for "house styles guide." Notice the shapes of the front, back, sides, and roof of the house style that you like best.

3. Using card stock, cut out pattern pieces for your house. You may need to do this multiple times to get the right shapes. For a very basic A-frame house, you can make 2 (3-by-3-inch) squares for the roof and 2 (3-by-3-by-2.5-inch) triangles for the front and back walls using the template on page 112.

4. Practice assembling your card stock house using tape to connect the card stock pieces. When you are satisfied with your design, remove the tape so that you have a stack of card stock pattern pieces.

5. Take the gingerbread dough out of the refrigerator and place it on the kitchen counter.

6. Preheat the oven to 350°F. Place a piece of parchment paper on each of the 2 cookie sheets. Sprinkle flour over a cutting board or freshly cleaned countertop.

7. Use the rolling pin to roll one of the gingerbread dough pieces until it is between ¼-inch and ½-inch thick. You may need to knead the dough in your hand for a few minutes to get it soft enough to roll smoothly.

INGREDIENTS:

FOR THE GINGERBREAD

- 5½ cups flour
- 1 teaspoon baking soda
- 1½ teaspoons salt
- 4 teaspoons ginger
- 4 teaspoons cinnamon
- 1½ teaspoons ground cloves
- 2 teaspoons nutmeg
- 1 cup (2 sticks) room-temperature unsalted butter
- 1 cup brown sugar
- 2 eggs
- 1½ cups molasses

FOR THE ROYAL ICING

- 1 pound confectioners' sugar
- 3 pasteurized egg whites, at room temperature
- ½ teaspoon cream of tartar

FOR DECORATING

- Brightly colored candy: gum drops, fruit snacks, peppermint swirls

CONTINUED

Hypothesis: Predict which shapes will make the most stable roof and walls for a gingerbread house.

_____

_____

_____

_____

Observations: What changes did you have to make to engineer a working design?

_____

_____

_____

_____

Results: What shapes did you choose for your final house design?

_____

_____

_____

_____

_____

_____

8. Lay your card stock pieces over the gingerbread and, with adult supervision, trace around each pattern piece with a knife, cutting out one gingerbread piece for each pattern piece.

9. Carefully place each gingerbread cutout on the parchment on a cookie sheet. Bake for 10 to 14 minutes, or until the gingerbread is firm on top but not cracked or hard.

10. While the gingerbread bakes, make the royal icing by mixing 1 pound of confectioners' sugar, 3 room-temperature pasteurized egg whites, and ½ teaspoon of cream of tartar on high speed for 7 to 10 minutes, until the icing keeps its shape when you turn off and remove the mixer.

11. Use the icing to connect the gingerbread pieces according to your pattern.

12. Use icing to glue the candy decorations to your engineered gingerbread home.

GINGERBREAD RECIPE

1. In one large mixing bowl, stir together 5½ cups of flour, 1 teaspoon of baking soda, 1½ teaspoons of salt, 4 teaspoons of ginger, 4 teaspoons of cinnamon, 1½ teaspoons of ground cloves, and 2 teaspoons of nutmeg.

2. In the other large mixing bowl, use the electric mixer to mix 1 cup (2 sticks) of room-temperature unsalted butter with 1 cup of brown sugar.

3. When the butter and sugar are well combined, add 2 eggs and 1½ cups of molasses. Mix on medium speed until well combined.

4. Using the rubber spatula, divide the dough into 3 pieces. Wrap each piece tightly in plastic wrap and refrigerate overnight.

The Hows and Whys Triangles and rectangles make great load-bearing walls because they distribute their stress evenly along their straight lines. Math helps us understand how different shapes handle stress.

STEAM CONNECTION: The design process is the key to successful engineering. When you created the card stock prototype of your house, you were deep into design-thinking mode. Every time your prototype fell down, you were that much closer to finding a design that worked. You kept trying even when you failed. That is what makes you an awesome engineer!

Turn It Up! When researching architectural styles, let your imagination run wild! Why not add a chimney or even a front porch to your gingerbread house? Use your card stock to make and remake shapes until you get a set that fits together to make your dream house. Don't give up! Every time you have to make a change, your house design gets stronger and better.

CONTINUED

GINGERBREAD HOUSE TEMPLATE

Roof (make 2)

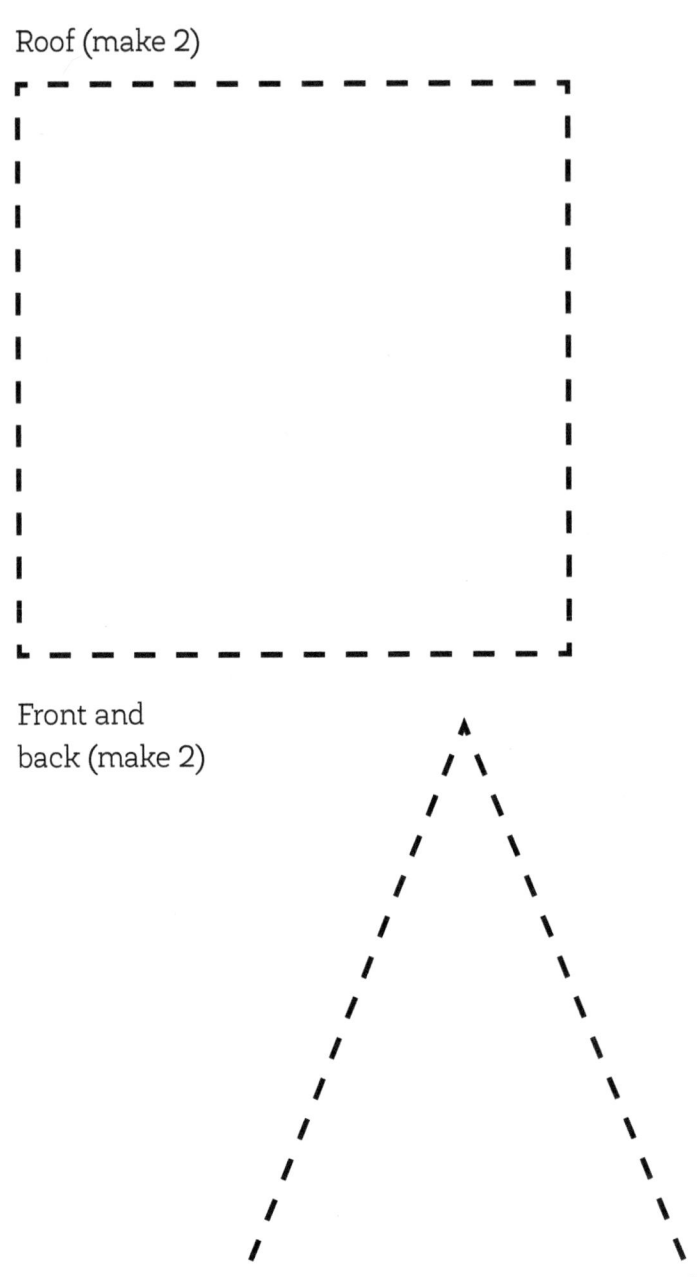

Front and
back (make 2)

EXTRA FEATURE: A SLICE OF HISTORY WITH GEORGE WASHINGTON CARVER

George Washington Carver is famous for inventing a food he didn't actually invent: peanut butter. Even though he didn't invent peanut butter, Carver's work as a scientist of **agriculture**—the science of growing food—made peanuts one of the most popular foods in America.

George Washington Carver was born into slavery during the American Civil War. Slavery ended with the war in 1865. George Washington Carver was a curious and hardworking kid, and eventually he was the first black student to attend Iowa State University, where he graduated with a master's degree in plant science.

As a scientist, Carver studied the plants that people grow for food. He noticed that when farmers grew the same crops in the same soil for many years, the soil became tired and the crops stopped growing. Carver engineered a solution. He taught farmers to plant different crops from year to year. He discovered that certain plants would make the soil healthy again. One of these plants was the peanut. To inspire farmers to grow peanuts and heal their soil, George Washington Carver invented over 300 peanut products, including peanut soap and peanut milk. Carver's work brought peanuts to American farms, and Americans fell in love with peanut butter (invented by Marcellus Gilmore Edson, John Harvey Kellogg, and Ambrose Straub). Every time you enjoy the sweet, salty taste of peanut butter, you can also enjoy knowing that the peanuts grew on a farm with rich, healthy soil.

Chapter Six

When food scientists use art to create their meals, things get interesting. Food can become brightly colored and beautifully decorated in the hands of a creative chef.

In this chapter, you will learn about the art of food science with a wild collection of experiments. The first two experiments focus on making candy. You will work with sugar, melting it at different temperatures and mixing in different ingredients. The next four experiments are about coloring food with dyes. You will explore color mixing and how to make natural dyes from vegetables. Next you will learn how to **dissect** a fruit, flower, and **fungus** as you prepare edible decorations. You'll learn more about food decoration by making potatoes look like people and desserts that resemble Earth's landforms.

As you work, pay attention to how you can change the texture and colors of foods. Understanding how temperature changes sugar is a powerful skill that could help you become an expert candy maker. Knowing how to make food coloring from vegetables can ramp up your food decorating game. Most of all, enjoy your delicious—and creative—dishes!

REAL ROCK CANDY:
CHOCOLATE-COATED CANDY GEODES

TOOLS:

- Saucepan
- Measuring cups
- Spoon
- Stove
- Silicone candy molds
- Microwave oven
- Plate
- 2 bowls

INGREDIENTS:

- 1 cup sugar
- ⅓ cup water
- Food coloring
- 1 cup white chocolate chips
- 1 cup milk chocolate or dark chocolate chips

LEVEL OF DIFFICULTY: ADVANCED
MESS-O-METER RATING: MEDIUM MESS
BEST EATEN FOR: DESSERT
PREP TIME: NONE
FULL TIME: 30 MINUTES TO PREPARE THE SUGAR SOLUTION, 2 TO 5 DAYS FOR THE CRYSTALS TO FORM, AND 30 MINUTES TO COAT THE GEODES WITH CHOCOLATE
MAKES: 10 LARGE CANDIES

 When beautiful rocks like geodes form in nature, they start out as liquid rock. As the rock cools, it forms **crystals**, or sparkling rocks. In this lab, you will make rock candy. What do you think the crystals will look like? Will you be able to see individual crystals? What shapes will they have?

Cautions: Adult supervision is needed when you are using the stove and when you are handling hot melted chocolate.

THE STEPS

1. Pour 1 cup of sugar and ⅓ cup of water in a saucepan.

2. Over low-to-medium heat, dissolve the sugar by stirring it in the water as it heats. This will get boring quickly, but don't stop stirring! It will take 5 to 20 minutes for

the sugar to fully dissolve. You will know it has dissolved when you can't see any gritty crystals of sugar in the solution (there may be a few grains of sugar on the sides of the pan).

3. Turn off the heat and let the hot sugar solution cool for a few moments.

4. Stir 5 drops of food coloring into the hot sugar solution.

5. Pour the hot sugar solution into the candy molds.

6. Let the candy molds sit at room temperature for 2 to 3 days so that crystals can start to form. Once a day, poke the crust on top of the hardening candy with a clean finger. This will break the crust and help speed up the drying process.

7. When the candy has hardened on the bottom, you will be able to pop it out of the silicone candy mold. Some of the pieces of candy might be a little bit wet with sugar solution. Set the pieces of candy on a clean plate to finish drying. Make your observations.

8. When the candy pieces are dry, pour 1 cup of white chocolate chips in a bowl and microwave for 30 seconds. Stir with a spoon. If the chocolate has not completely melted, microwave for 10 seconds and stir. Repeat until the chocolate is melted.

9. Dip each piece of candy in the white chocolate. Set the candy to dry on a clean plate.

10. When it's dry, your candy is finished! You can crack a geode open by hitting it with the back of a spoon. Record your results.

CONTINUED

Hypothesis: Predict what the sugar crystals will look like in your rock candy geodes.

_____

_____

_____

_____

_____

Observations: Describe how your rock candy geodes look when you first take them out of the silicone molds.

_____

_____

_____

_____

_____

Results: Did the crystals in your rock candy geodes look the way you expected them to look?

_____

_____

_____

_____

_____

The Hows and Whys Rock candy geodes form the same way that actual geodes form. When some liquids (including sugar solution and melted agate, amethyst, and opal) cool, the molecules make a pattern. Crystals are made out of very organized patterns of solid molecules.

STEAM CONNECTION: Geologists are Earth scientists who study rocks. Understanding how rocks form helps geologists find and identify geodes and other precious gems.

Turn It Up! Look up different types of geodes online. Try using different colors and layers of chocolate to make rock candy geodes that look like a variety of types of geodes.

STAINED GLASS WONDERS:
COLOR SWIRL CANDY

LEVEL OF DIFFICULTY: ADVANCED
MESS-O-METER RATING: MEDIUM MESS
BEST EATEN FOR: DESSERT
PREP TIME: NONE
FULL TIME: 2 HOURS
MAKES: ABOUT 4 CUPS OF HARD CANDY

Food scientists add different ingredients to sugar to get different types of candy. In this experiment, you will make glass candy using two recipes. Do you think that your candy glass will be smoother if you use just sugar and water, or if you add other ingredients?

Cautions: Adult supervision is needed for this entire experiment. Be sure to wear oven mitts and an apron whenever handling the hot sugar. When you pour the hot sugar into a pan, the pan will become very hot very quickly. Stay clear of the hot sugar until it cools. Be careful of any sharp edges on your finished candy. To clean up candy drips, soak with water and they will dissolve within 5 minutes.

TOOLS:

- 2 (9-by-13-inch) cake pans
- Measuring cups and spoons
- Pot
- Candy thermometer with clip
- Long-handled spoon
- Stove
- Oven mitts
- Table knife

INGREDIENTS:

- Oil or cooking spray
- 2 cups water, divided
- 3½ cups sugar, divided
- Food coloring (regular or gel)
- ½ cup corn syrup
- ⅛ teaspoon cream of tartar

CONTINUED

Hypothesis: Will a solution of sugar and water make a smoother candy glass than a solution of sugar, water, corn syrup, and cream of tartar? Why?

_____

_____

_____

_____

_____

Observations: Describe the appearance of the sugar solutions as you poured them into the pans to cool.

_____

_____

_____

_____

_____

Results: Which recipe made the smoother candy glass?

_____

_____

_____

_____

_____

THE STEPS

1. Coat the bottom of 2 (9-by-13-inch) cake pans with oil or cooking spray.

2. Pour 1 cup of water and 1¾ cups of sugar into the pot. Clip the candy thermometer to the side of the pot.

3. Stir the sugar solution with the long-handled spoon.

4. Stirring constantly, heat the sugar solution over low heat until it begins to boil. Watching the temperature carefully, turn the heat up to medium. Continue stirring the sugar solution.

5. When the sugar solution reaches 300°F to 310°F, turn off the stove. Pour the sugar solution quickly into one of the cake pans. Record your observations.

6. Carefully add 10 to 20 drops food coloring across the top of the sugar solution.

7. Use a table knife to swirl the colors a little, but do not mix the sugar solution.

8. Leave the sugar solution for 1 hour to cool.

9. Repeat steps 2 to 8, but this time add ½ cup of corn syrup and ⅛ teaspoon of cream of tartar to the 1 cup of water and 1¾ cups of sugar in step 2.

10. Set both cake pans upside-down over a clean counter. Tap the pans until the candy falls out. It will break when it hits the counter. The candy is ready to eat! Record your results as you enjoy your dessert.

The Hows and Whys When sugar cools, it usually forms crystals that sparkle. This sugar candy is not smooth. Corn syrup and cream of tartar create a chemical change with sugar that stops crystals from forming so that the solid sugar is smooth like glass.

STEAM CONNECTION: Sugar science is a combination of biology, chemistry, and physics. Sugar comes from living things like sugar cane or sugar beets, and biology helps scientists understand how sugar is made. Sugar is a chemical, and chemistry helps scientists understand the kinds of chemical changes that sugar can undergo. Sugar crystals reflect light, and physics helps scientists understand how sugar sparkles or shines.

Turn It Up! There are so many different sweets you can make with plain sugar. Check out the Exploratorium website listed in the Resources section for a full list, then try making one of the recipes listed there—like fudge, caramels, or taffy.

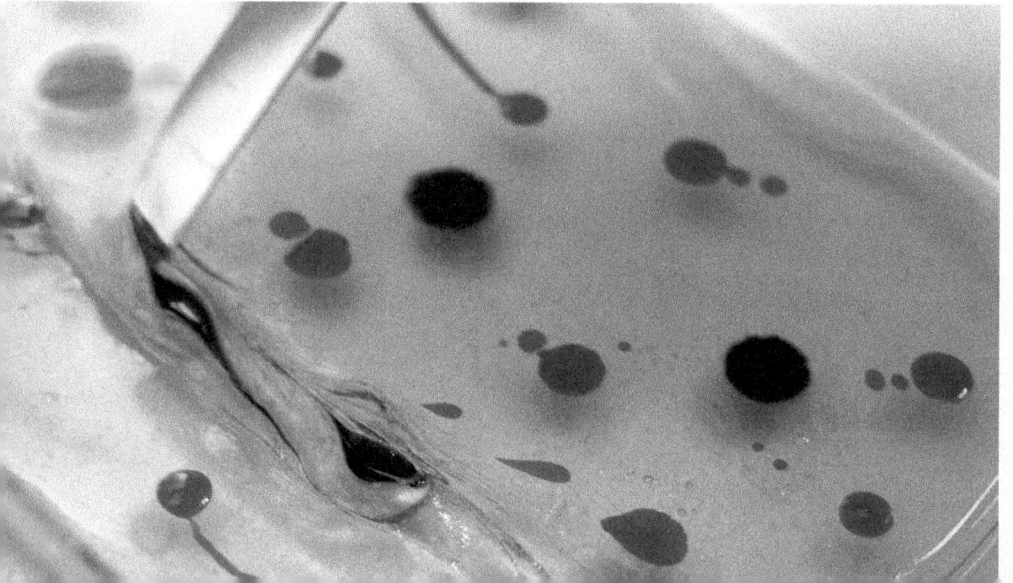

OVER THE RAINBOW: COLOR MIX FROSTING

TOOLS:

- 2 bowls
- 2 spoons

FOR THE FROSTING

- Measuring cups and spoons
- Large mixing bowl
- Electric or stand mixer

INGREDIENTS:

- About 1½ cups frosting (see recipe on page 123 or purchase premade frosting)
- Regular food color
- Gel food color

FOR THE FROSTING

- ½ cup unsalted butter, at room temperature
- 1½ cups powdered sugar
- 2 tablespoons heavy whipping cream
- 1 teaspoon flavored extract (optional)

LEVEL OF DIFFICULTY: EASY
MESS-O-METER RATING: MINOR MESS
BEST EATEN FOR: DESSERT
PREP TIME: 10 MINUTES TO MAKE FROSTING
FULL TIME: 15 MINUTES
MAKES: ABOUT 1 CUP FROSTING

? Cake decorators study different types of food coloring to discover how to bring their artistic frosting designs to life. In this experiment, you will test different types of food coloring to discover: Will regular food color or gel food color make brighter frosting?

! **Cautions:** If making frosting from scratch, have an adult run the electric mixer.

THE STEPS

1. Put half of your frosting in one bowl, and the other half of your frosting in a second bowl.

2. Choose one color: green, orange, or purple. Pick out the food colors that you will need to make this color.

   a. Green: Use blue and yellow
   b. Orange: Use red and yellow
   c. Purple: Use red and blue

3. Using the regular food color, add 2 drops of each color needed to the first bowl of frosting. Stir with a spoon. Record your observations.

4. Using the gel food color, add 2 drops of each color needed to the second bowl of frosting. Stir with the second spoon. Record your observations.

5. Record your results.

BASIC BUTTERCREAM FROSTING

1. Using an electric mixer, beat ½ cup of room-temperature butter for 1 minute or until smooth.

2. Add 1½ cups of powdered sugar. Beat at low speed for 30 seconds and at medium speed for 1 minute or until creamy.

3. Add 2 tablespoons of heavy whipping cream and mix until creamy.

4. Optional: Add 1 teaspoon of your favorite flavor extract (like vanilla, almond, or lemon).

CONTINUED

Hypothesis: Predict whether regular food color or gel food color will make brighter frosting.

_____

_____

_____

_____

Observations: What do you notice when you mix the two types of food color into the frosting?

_____

_____

_____

_____

Results: Which type of food color made the brighter frosting?

_____

_____

_____

_____

5. Note: You can store this frosting in the refrigerator for a few days or in the freezer for up to 2 months. The frosting needs to be at room temperature, or a little bit warmer (but not melted), when you use it to decorate. For the easiest experience when frosting cakes or cupcakes, pop the cake in the freezer for 30 minutes ahead of time so that your frosting is much warmer than your cake.

The Hows and Whys Regular food color is water-based and can't always dissolve in buttery frosting. Gel food colors have molecules that help the colors dissolve in buttery frosting, so they often look brighter.

STEAM CONNECTION: Food scientists working to make creative decorations, including frosting on cakes and cupcakes, use color science in their designs and planning. Color science is a part of **optics**, a branch of physics that focuses on light. Color **psychologists** explore how different colors make people feel.

Turn It Up! A **color wheel** is a diagram that artists use to organize **primary** and **secondary colors**. For a more advanced experiment, make a double batch of buttercream frosting. Make at least six different colors of frosting to create a color wheel. Look up color wheels online and use the images you find as a guide. Use your frosting to decorate a round cake and you will have an edible color wheel!

DYEING FOR PASTA: DOES SHAPE MAKE A DIFFERENCE?

LEVEL OF DIFFICULTY: EASY

MESS-O-METER RATING: MINOR MESS

BEST EATEN FOR: LUNCH OR DINNER

PREP TIME: 15 MINUTES TO COOK THE PASTA

FULL TIME: 30 MINUTES

MAKES: 2 BOWLS OF PASTA

? Food dyes make it possible to create artistic meals, but they don't always work well. In this experiment, you will dye curly and straight noodles to test whether the shape of pasta matters for food dyes. Will food dye work better, worse, or the same for curly noodles versus straight noodles?

CONTINUED

TOOLS:

- Pot
- 2 bowls
- 2 spoons
- Stove
- Colander

INGREDIENTS:

- Water
- ¼ pound curly pasta, cooked
- ¼ pound straight pasta, cooked
- Food dye (regular or gel)

Hypothesis: Predict whether food dye will work better, worse, or the same on curly noodles or straight noodles.

_____

_____

_____

_____

_____

Observations: Describe what you see when you place the cooked noodles into the bowls of dye.

_____

_____

_____

_____

_____

Results: Did food dye work better, worse, or the same on curly noodles or straight noodles?

_____

_____

_____

_____

 Cautions: Have an adult prepare the pasta by boiling it in water according to the directions on the package.

THE STEPS

1. Have an adult prepare ¼ pound (1 cup) of curly pasta and ¼ pound (1 cup) of straight pasta by boiling each in water according to the directions on the package.

2. While the pasta cooks, pour ½ cup of warm water and 4 drops of food coloring in each of the 2 bowls.

3. When the pasta is cooked and drained, put the curly pasta in one bowl and the straight pasta in the other bowl.

4. Use the spoons to stir the pastas in their bowls. Record your observations and results.

5. Drain the pastas in a colander. Rinse with warm water. Top with your favorite sauce (or butter) and enjoy.

The Hows and Whys Food dye is a pigment that sticks to food molecules. This is a chemical change. Food dye pigment sticks to some food molecules better than others, but this is because of the shapes of the molecules, not because of the shape of the food. Pasta is made of the same molecules, no matter what shape it is in—so all shapes of pasta work the same with food dye.

STEAM CONNECTION: Understanding the chemical changes involved in food has helped scientists develop the rainbow of food dyes that we use today. While artificial dyes made in laboratories and factories are still common, there are more natural dyes being used today than there were in the past.

Turn It Up! Try using food dyes (use store-bought dye or try the Beet It experiment on page 128 to make your own dye) on grains other than pasta, like rice, potatoes, or bread. What grains dye the most easily? Does adding dye to food make it look better to eat?

BEET IT: EDIBLE DYES

TOOLS:

- 2 small cooking pots
- Liquid measuring cup
- Stove

INGREDIENTS:

- 4 cups water, divided
- 1 cup vinegar
- 2 beets, sliced
- 2 hard-boiled eggs, peeled
- Pinch salt

Hypothesis: Predict whether making beet dye with vinegar for a mordant will make a brighter dye.

_____

_____

_____

_____

_____

LEVEL OF DIFFICULTY: MEDIUM

MESS-O-METER RATING: MEDIUM MESS

BEST EATEN FOR: LUNCH OR SNACK

PREP TIME: 20 MINUTES FOR AN ADULT TO MAKE HARD-BOILED EGGS AND SLICE THE BEETS

FULL TIME: 1 HOUR

MAKES: 2 SERVINGS OF PICKLED EGGS WITH BEETS

? Natural dyes are a great alternative when you want brightly colored healthy food. Natural dyes use a **mordant**, or chemical that helps the dye work. In this experiment, you will make a natural dye using beets for color and vinegar for a mordant. How important is a mordant for making bright colors?

! Cautions: Adult supervision is needed when cooking on the stove. Have an adult slice the beets and the eggs. Beet juice will stain. Wear an apron to protect your clothes.

THE STEPS

1. To hard-boil eggs, place them in boiling water for 14 minutes. Remove the eggs from the hot water with a slotted spoon and put them in a bowl of cold water.

2. Pour 2 cups of water into a pot and add 1 sliced beet.

3. Pour 2 cups of water, 1 cup vinegar, and 1 sliced beet into another pot.

4. Heat both pots over high heat until their contents boil.

5. Turn the heat to low and simmer for 15 minutes.

6. Turn off the heat and let the pots cool off for 15 minutes.

7. Peel the eggs and place one into each pot of cooled beet soup. Let the eggs soak for 10 minutes.

8. Remove the eggs. Have an adult cut each egg in half. Record your observations and results.

9. Put the cooked, sliced beets on a plate with the sliced eggs. Sprinkle the eggs with a pinch of salt. Pickled eggs and beets are a dish traditional to the American South. Give them a try!

The Hows and Whys Some plants, including beets, make very strong pigments. Even strong color molecules stick better when we use a mordant to help the dye set.

STEAM CONNECTION: Understanding fiber science helps us use natural dyes successfully. Wool takes natural dyes well, and fiber artists often dye wool yarn with plant-based dyes. You can tie-dye cotton fabric, including T-shirts, with natural dyes.

Turn It Up! Make dye from other plants, herbs, and spices. Kale makes green dye, red cabbage makes purple dye, carrots make orange dye, turmeric, a bright yellow spice, makes yellow dye, and rosemary makes light yellow dye.

Observations: Is there any difference in appearance between the eggs dyed with the vinegar mordant?

_____

_____

_____

_____

_____

Results: Was a mordant needed for dyeing eggs with beets?

_____

_____

_____

_____

CRUNCHY CAPILLARY ACTION:
COLOR CHANGE CELERY

LEVEL OF DIFFICULTY: EASY
MESS-O-METER RATING: MINOR MESS
BEST EATEN FOR: SNACK
PREP TIME: NONE
FULL TIME: 35 MINUTES TO SET UP, OVERNIGHT TO RUN, AND 5 MINUTES TO OBSERVE THE NEXT DAY
MAKES: 2 CRUNCHY CELERY STICKS

TOOLS:

- 2 tall, clear glasses
- 1 knife for an adult to cut the celery

INGREDIENTS:

- Food coloring (regular or gel)
- 2 long stalks of celery

 Water is the most important molecule in living things. On a hot day, plants lose water through their leaves. They make up for the lost water by taking water from the ground through their roots. The water moves up through plants—defying **gravity**! In this experiment, you will observe this amazing water movement, called **capillary action**. Do you think capillary action works better in plants that are dry or plants that are wet?

Cautions: Have an adult use the knife to cut the celery.

THE STEPS

1. Fill 2 tall, clear glasses with water.

2. Put 4 drops of food coloring in each glass.

3. Wash 2 long celery sticks.

Hypothesis: Predict whether you will see more capillary action in wet or dry celery.

_____

_____

_____

_____

_____

Observations: Describe what you see after the celery has soaked overnight.

_____

_____

_____

_____

_____

Results: Did the wet celery or the dry celery have more capillary action?

_____

_____

_____

_____

_____

4. Have an adult use a knife to cut the bottom 2 inches and the top 3 inches off each celery stick.

5. Place 1 celery stick into a glass of color water immediately. There should be at least 1 inch of celery sticking up above the water.

6. Set the other celery stick on the counter to dry for 30 minutes.

CONTINUED

7. Place the second celery stick into the second glass of color water. There should be at least 1 inch of celery sticking up above the water.

8. Let both celery sticks soak overnight.

9. Record your observations and results.

The Hows and Whys In the wet celery stalk, water evaporated out of the top of the celery. New, colored water came in from the glass to replace the lost water. Unlike most chemicals, water sticks to itself well. The water that was already in the celery stuck to the colored water in the glass and helped pull it up using capillary action. Dry celery has bubbles of air in the water tubes. The bubbles of air stop the capillary action.

STEAM CONNECTION: Capillary action is a big part of agriculture. Agricultural scientists and engineers have to understand how to keep plants watered in order for food crops to grow.

Turn It Up! Capillary action can also be used to add food coloring to flowers. Find a pesticide-free white rose and soak it in a mixture of food color and water overnight. You can eat the colorful petals on a salad the next day!

INSIDE THE ROSE:
FRUIT, FLOWER & FUNGUS DISSECTIONS

LEVEL OF DIFFICULTY: EASY

MESS-O-METER RATING: MINOR MESS

BEST EATEN FOR: SNACK OR GARNISH

PREP TIME: 5 MINUTES FOR AN ADULT TO SLICE THE FRUIT, FLOWER, AND MUSHROOM

FULL TIME: 30 MINUTES

MAKES: 1 SMALL SNACK OR SEVERAL GARNISHES

Scientists who study biology often dissect, or cut open, living things. In this experiment, you will dissect a tomato, a flower, and a mushroom, which is a kind of fungus. What do you think will be the same or different inside a fruit, a flower, and a fungus?

Cautions: Have an adult slice the foods. Be sure that your edible flowers are pesticide-free. Flowers sold for decorations often have very high amounts of pesticides on them.

THE STEPS

1. Have an adult slice a tomato, an edible flower, and a mushroom in half.

2. Place one half of the tomato on a clean plate. Sit down at a table with the tomato, drawing paper, and coloring tools. Make a detailed drawing of what you see.

CONTINUED

TOOLS:

- Knife and cutting board
- Drawing paper and coloring tools

INGREDIENTS:

- 1 large tomato
- 1 or more edible flowers, such as nasturtiums, pansies, marigolds, borage, tulips, or roses
- 1 large mushroom
- 1 tablespoon of salad dressing

Hypothesis: Predict what you will see inside the fruit, flower, and fungus.

_____

_____

_____

_____

_____

Observations: Make detailed drawings of each food. Describe what you see inside the fruit, flower, and fungus. If you are not sure what you are seeing, do an online search for "tomato cross-section diagram," "flower cross-section diagram," or "mushroom cross-section diagram."

_____

_____

_____

_____

_____

_____

_____

_____

_____

Results: What was the same inside each food? What was different?

_____

_____

_____

_____

_____

3. Repeat with the edible flower and then with the mushroom. Record your observations.

4. Record your results.

5. If the pieces of your tomato or mushroom are very large, have an adult slice them into bite-size pieces. Put everything on one clean plate or use the slices and flowers to decorate the next meal you eat as a garnish.

The Hows and Whys Fruits and flowers are both parts of plants. As a plant moves through its life cycle, its flowers turn into fruits with seeds. You may have noticed that fruits and flowers have several parts in common. Mushrooms are not plants, so they look different from fruits and flowers inside.

STEAM CONNECTION: Careful observations made through drawing as well as writing—two important art forms—help scientists keep track of their experiments and inventions. Learning to draw carefully and slowly will make you a better observer.

Turn It Up! Botanical drawings are an old art form. Check out a book of botanical illustrations from your local library and try creating more of your own. Or plant some nasturtium or borage seeds in your garden. They will grow lots of flowers for you to eat!

AN A-PEELING DISH:
POTATO PEOPLE

LEVEL OF DIFFICULTY: MEDIUM
MESS-O-METER RATING: MINOR MESS
BEST EATEN FOR: LUNCH OR DINNER
PREP TIME: NONE
FULL TIME: 1 HOUR 40 MINUTES (10 MINUTES TO PREPARE THE POTATOES, ABOUT 1 HOUR TO BAKE, 15 MINUTES TO COOL, AND 15 MINUTES TO DECORATE)
MAKES: 4 LARGE POTATOES

 When food scientists study heat, they often try to find techniques that speed up baking time. Will plain potatoes or potatoes wrapped in aluminum foil bake faster?

Cautions: Adult supervision is needed whenever you use the oven.

THE STEPS

1. Preheat the oven to 400°F.

2. Wash and dry 4 potatoes and carefully prick them with a fork. (This prevents the potatoes from exploding in the oven.)

3. Wrap 2 of the potatoes in aluminum foil.

4. Place all potatoes on the cookie sheet.

5. When the oven is at 400°F, place the cookie sheet into the oven. Set the kitchen timer for 30 minutes.

CONTINUED

TOOLS:

- Oven
- Fork
- Aluminum foil
- Cookie sheet
- Kitchen timer
- Oven mitts
- Plate

INGREDIENTS:

- 4 large russet potatoes
- 1 cup shredded cheddar or marble jack cheese
- Decorations, such as:
  - Olives
  - Yellow, red, or green pepper, cut into thin strips
  - Pea pods
  - Cherry or grape tomatoes
  - Baby corn
  - Baby carrots
  - Broccoli florets
  - Spinach leaves
  - Sliced mushrooms
  - Sliced cucumbers

Hypothesis: Predict whether potatoes will bake faster if they are wrapped in aluminum foil or baked plain.

————————————
————————————
————————————
————————————
————————————

Observations: Write down how many minutes it took each of your 4 potatoes to bake and make a note beside each time so you know if the potato was wrapped in aluminum foil or not.

————————————
————————————
————————————
————————————
————————————

Results: Which baking method was faster?

————————————
————————————
————————————
————————————

6. When the timer goes off, test the potatoes to see if they are finished baking by sticking them with the fork. When the fork slides all the way through a potato easily, the potato is cooked. When a potato is fully cooked, use an oven mitt to take it out of the oven and set it on a plate to cool.

7. If the potatoes are not done yet, set the timer for 10 minutes and check them again when it goes off.

8. Repeat step 7 until all 4 potatoes are done baking. Record your observations and results.

9. When the potatoes are cool enough to handle safely, have an adult slice each potato lengthwise so that you have 8 long half-potatoes.

10. Use the decorations and the cheese to make each half-potato into a face. Make eyes, eyebrows, mouths, noses, ears, and hair—go all out! Be sure to use plenty of cheese.

11. Enjoy your potatoes with your family.

The Hows and Whys Aluminum is a metal, and metals **conduct**, or move, heat. When you wrap a potato in aluminum foil, the potato bakes faster because the foil traps heat. However, the foil also traps water and steam around the potato, so its skin won't get as crispy as the skin on the potato baked without foil. It's up to you to decide if the crispy potato skin is a-*peeling* enough to justify the extra baking time.

STEAM CONNECTION: Baking time is a big deal for food scientists. Convection ovens are a technology developed to bake foods faster. Convection ovens have a fan inside that moves the hot air to evenly heat and cook food.

Turn It Up! Test your baking methods on different types of potatoes. Try sweet potatoes, yams, or Yukon Golds.

DELICIOUS DINOSAURS: MODELING ROCK & FOSSIL FORMATION

LEVEL OF DIFFICULTY: ADVANCED
MESS-O-METER RATING: MEDIUM MESS
BEST EATEN FOR: DESSERT
PREP TIME: 45 MINUTES TO MAKE THE BOX CAKE MIX
FULL TIME: 10 MINUTES TO MAKE THE FOSSILS, 1 HOUR FOR THE FOSSILS TO HARDEN, AND 20 MINUTES FOR THE MODELING ACTIVITY
MAKES: 1 SCRUMPTIOUS CHOCOLATE CAKE

? **Fossils** are reminders of living things from Earth's long-lost history. But how do fossils get here? In this experiment, you will use a scientific process called **modeling** to go through the steps of making fossils and the rocks that hold them. You will make three kinds of rocks: **igneous** (formed when molten rock such as **lava** cools), **metamorphic** (rock that is heated but does not melt completely), and **sedimentary** (made up of layers of dirt and sand). In which kind of rock do you think fossils will form?

! **Cautions:** Adult supervision is required when you bake the cake and melt the chocolate. Melted chocolate is very hot and should not be touched.

TOOLS:

- Liquid measuring cup
- Measuring cups and spoons
- Bowl
- 2 spoons
- 9-by-13-inch cake pan
- Oven
- Oven mitts
- Small cake pan (for example, 5-by-9 inches) or pie pan
- Dinosaur toys, washed and dried
- Microwave oven
- Table knife

CONTINUED

INGREDIENTS:

- ➤ 1 box chocolate cake mix
- ➤ Eggs, oil, and water to make the cake mix
- ➤ 1 (2-pound) bag of brown sugar
- ➤ 1 (2-cup) bag of white chocolate chips
- ➤ 1 can prepared chocolate frosting
- ➤ 10 mint leaves
- ➤ 10 Oreo® cookies

Hypothesis: How do you think fossils form?

_____

_____

_____

_____

_____

_____

_____

_____

THE STEPS

1. Prepare the chocolate cake mix according to the directions on the box. Let the cake cool to room temperature after it bakes.

2. Pour the entire package of brown sugar into a small cake pan or pie pan. Use the back of a spoon to pat it down.

3. Arrange as many dinosaur toys as you can fit over the brown sugar. Gently press each toy down to make a dinosaur-shaped dent in the brown sugar. Remove the dinosaur toys.

4. Pour 2 cups of white chocolate chips into a heat-proof bowl. Microwave for 30 seconds, and then stir. Continue microwaving for 10 seconds at a time, stirring after each time, until the white chocolate is completely melted.

5. Slowly pour the melted white chocolate into the dinosaur-shaped dents in the brown sugar. Place the pan in the refrigerator for 1 hour.

6. When the white chocolate has hardened, carefully dig each fossil out of the brown sugar with a spoon. Brush the brown sugar off with your clean fingers. Place the fossils on top of the cooled chocolate cake so that they are spread out across the whole cake.

7. Open the can of prepared frosting and remove the foil seal. Microwave the frosting for about 20 seconds, until it is very easy to stir and pour but not completely melted.

8. Pour the warm frosting over the cake, filling in the spaces between the fossils. Use a table knife to smooth the frosting over the entire cake, covering the fossils.

9. Let the frosting cool for 10 minutes.

10. Use a clean dinosaur toy to make dinosaur footprints across the top of the frosting.

11. Press the mint leaves into the frosting.

12. Crush 10 Oreo cookies and sprinkle the crumbs evenly over the cake.

13. Record your observations and results.

> **The Hows and Whys** In this model, the cake batter represents molten rock (lava or **magma**) and the baked cake is igneous rock. The almost-melted frosting cools to form metamorphic rock. The cookie crumbs are loose pieces of dirt and sand that, over many years, form sedimentary rock. Fossils form in sedimentary rock.

STEAM CONNECTION: Scientists and engineers use models to understand and predict all kinds of future events and their effects. For example, scientists build computer and mathematical models to predict the impact of **global climate change**.

Turn It Up! Look up **molds**, **casts**, **trace fossils**, and **preserved remains** in the Glossary. What parts of your cake modeled each of these kinds of fossils?

Observations: Take notes on how igneous, metamorphic, and sedimentary rocks are made.

_____

_____

_____

_____

_____

_____

_____

_____

_____

_____

_____

_____

Results: How did this activity model fossil and rock formation?

_____

_____

_____

_____

_____

_____

_____

_____

YOUR OWN PERSONAL ICE CREAM AGE: MODELING GLACIERS

LEVEL OF DIFFICULTY: EASY

MESS-O-METER RATING: MINOR MESS

BEST EATEN FOR: DESSERT

PREP TIME: NONE

FULL TIME: 15 MINUTES

MAKES: 4 SERVINGS OF ICE CREAM WITH MIX-INS

At least five times in Earth's history, the whole planet has plunged into an **Ice Age**: a time when global temperatures are cold enough for huge rivers of ice, or **glaciers**, to form. When they move across the land, glaciers scrape off dirt and rocks. The dirt and rocks can mix into the glacier as it moves. In this experiment, you will model the movement of a glacier using ice cream. How much dirt and rocks (mix-ins) do you think your glacier will pick up?

THE STEPS

1. Place the wafer cookies in a single layer in your baking pan, covering the entire bottom. This is the **bedrock**, or solid rock, underneath the dirt and loose rocks.

2. Sprinkle your mix-ins on top of the wafer cookies. You want a nice, big heap of dirt and rocks.

3. Place a book or other object under one end of the pan so that it is like a hill.

CONTINUED

TOOLS:

- 9-by-13-inch baking pan
- A book or other 2- to 4-inch-tall object to lift up one end of the pan
- Ice cream scoop
- 4 spoons
- 4 bowls (optional)

INGREDIENTS:

- 1 package vanilla or chocolate wafer cookies
- 1 pint of your favorite flavor of ice cream
- At least 4 types of crumbly ice cream mix-ins:
  - Crushed Oreos, chocolate chip cookies or graham crackers
  - M&M's™
  - Chocolate chips
  - Mini marshmallows
  - Sprinkles
  - Shredded coconut

Hypothesis: Predict what portion of the dirt and rocks will be picked up by your glacier: ¼, ½, ¾, or all of them.

_____

_____

_____

_____

_____

_____

Observations: How did the glacier pick up dirt and rocks?

_____

_____

_____

_____

_____

Results: About how much of the dirt and rocks went into the glacier? Was this what you were expecting?

_____

_____

_____

_____

4. Scoop 1 pint of ice cream on top of the mix-ins and wafer cookies at the high end of the hill (the pan).

5. Wait 5 minutes for the ice cream to start melting.

6. Your ice cream glacier should start sliding down the hill. If it needs a little push, gently scoot it with a spoon.

7. Record your observations and results.

8. Divide your glacier and mix-ins between 4 bowls or use the spoons to eat right out of the pan. Yum!

The Hows and Whys Glaciers are so heavy that the pressure from their weight melts a layer of water between the glacier and the land. This water lets the glacier slide across the land. As the glacier moves, it picks up entire landforms—even hills and mountains! Did your glacier pick up huge pieces of mix-ins?

STEAM CONNECTION: Events in nature that happen at a large scale, over long periods of time, or deep in Earth's past are almost impossible for Earth scientists to observe. Modeling is an important scientific process that lets us observe smaller versions of these events.

Turn It Up! How would you build a model of a river using food? The ScienceBuddies website listed in the Resources section has directions for building a model river with cornmeal, sand, and water.

EXTRA FEATURE: CAREERS IN FOOD SCIENCE

Professional food scientists take food preparation to the next level. Let's learn about a few of the thousands of opportunities for careers in food science!

Food companies have teams of food scientists inventing everything from turkey pot pie to candy bars. These food scientists are engineers who use the design process to create new food products and improve existing products.

Once these products are ready to sell, food scientists use hundreds of safety checks to make sure that the food these companies send to stores is safe. They want their products in packages that will look good to shoppers, and they want their trucks, trains, boats, and planes to deliver their products on time and in good condition.

Food scientists don't all work for big companies. Some of them work for governments or charity organizations to develop foods that will solve hunger problems. Food scientists also work outdoors with farmers to discover new, better ways to grow crops. All of these scientists have studied STEAM in school.

If you love cooking and science, you can become a food scientist, too!

Chapter Seven

MATHEMATICS

Imagine trying to make a batch of macaroni and cheese without knowing how much milk to add. You might get soupy pasta—yuck! Food scientists use math to create perfect recipes. In kitchen experiments, we use math every time we make a measurement, take a temperature, or calculate a cooking time. If math seems mysterious to you, never fear! Kitchen science math makes sense and tastes amazing.

In this chapter, you are going use math to make delicious recipes. You will explore density—which is calculated by dividing one number by another—with a delicious drink and fresh, crispy popcorn. You will explore **pH**—the mathematical measurement of how much acid or base is in a solution—with cabbage soup, lemonade, and chocolate cake (yum!). But that's not all. You get to learn about temperature measurements by making ice cream and calculating proportions in recipes with edible marshmallow slime. To top it off, you will learn about how temperature changes the volume of a gas—and measure this change—by baking a fluffy pancake.

LAYERS OF ARCHIMEDES:
DRINKABLE DENSITY COLUMN

TOOLS:

- 4 clear drinking glasses

INGREDIENTS:

- 1 (12-ounce) can bubbly water, any fruit flavor
- 1 prepared smoothie made with whole fruit
- 2 cups of clear fruit juice, like cranberry, pomegranate, apple, or berry

Hypothesis: Predict which of your three liquids will sink to the bottom of your glasses.

_____

_____

_____

_____

LEVEL OF DIFFICULTY: EASY
MESS-O-METER RATING: MINOR MESS
BEST EATEN FOR: SNACK
PREP TIME: NONE
FULL TIME: 10 MINUTES
MAKES: 4 DRINKS

Over two thousand years ago, a scientist named Archimedes discovered that when you put something in water, it moves the water and is lifted up by a force equal to the weight of the moved water. Mathematicians have been using **Archimedes' principle** to figure out the density of objects ever since. In this lab, you will observe three different liquids with three different densities—and then drink them! Which liquid do you think will sink to the bottom: bubbly water, smoothie, or juice?

THE STEPS

1. Set 4 clear drinking glasses on the kitchen counter.

2. Open 1 can of bubbly water. Slowly pour ¼ of the bubbly water into each glass. Record your observations.

3. Open 1 prepared smoothie. Slowly pour ¼ of the smoothie into each glass, letting the liquid run down the inner side of the glass. Record your observations.

4. Slowly pour ¼ of the 2 cups of juice into each glass, letting the liquid run down the inner side of the glass. Record your observations and results.

5. Share your drinkable **density column** with friends and family!

The Hows and Whys The liquids with the most density, or mass per volume, sink to the bottom of the column. The smoothie had the highest density and sank to the bottom because it had whole fruits, which have lots of mass, blended into it. The juice had the next highest density because the sugar in the juice added extra mass. The bubbly water was much less dense and floated on top because it had air bubbles, which have much less mass than sugar or fruit.

STEAM CONNECTION: Scientists often use math to calculate the density of the liquids they study.

Turn It Up! Explore the densities of other liquids—and solid foods—in your kitchen. Can you find a fruit that floats in between the smoothie and juice layers of your density column?

Observations: Describe what happens after you pour each liquid into the glasses.

_____

_____

_____

_____

_____

_____

_____

Results: Which liquid sank to the bottom? Which liquid floated to the top? Was this what you expected?

_____

_____

_____

_____

_____

_____

_____

_____

_____

_____

EXPLOSIVE PREDICTIONS: POPCORN ON THE FLY

TOOLS:

- ➲ Measuring cups and spoons
- ➲ Small pot (about 2 quarts) with lid
- ➲ Stove
- ➲ 4-cup liquid measuring cup, heat-proof
- ➲ Bowl
- ➲ Spoon

INGREDIENTS:

- ➲ 1 tablespoon canola or vegetable oil (not olive oil)
- ➲ ¼ cup popcorn kernels
- ➲ Pinch salt
- ➲ ⅛ cup grated cheddar cheese (optional)
- ➲ 2 teaspoons melted butter (optional)

LEVEL OF DIFFICULTY: MEDIUM
MESS-O-METER RATING: MINOR MESS
BEST EATEN FOR: SNACK
PREP TIME: NONE
FULL TIME: 15 MINUTES
MAKES: POPCORN FOR 1 TO 4 PEOPLE, DEPENDING ON HOW HUNGRY YOU ARE

 Popcorn is an amazing food. Like magic, kernels of popcorn explode when heated. In this experiment, you will predict the change in volume that happens when popcorn pops. How many cups of popcorn do you predict will pop out of ¼ cup of popcorn kernels?

Cautions: Adult supervision is needed when cooking on the stove.

THE STEPS

1. Pour 1 tablespoon of oil into a small pot (about 2 quarts) with a lid.

2. With adult supervision, heat the oil in the pot on medium heat for about 30 seconds.

3. Add ¼ cup of popcorn kernels to the pot and put the lid on.

4. Stand by the pot and listen. At first you will only hear one or two pops. Then there will be so many pops that you

AWESOME KITCHEN SCIENCE EXPERIMENTS FOR KIDS

Hypothesis: Predict how many cups of popcorn you will make when you start with ¼ cup popcorn kernels.

_____

_____

_____

_____

Observations: Describe what happens when you pop the popcorn.

_____

_____

_____

_____

can't count them. As soon as the pops start to slow down, turn off the stove burner.

5. Leave the pot with the lid on for about 1 minute while the popping stops.

6. With adult supervision, use an oven mitt to take the lid off the pot. Record your observations.

7. Use an oven mitt to lift the pot and pour the popcorn into a 4-cup measuring cup. Record your results.

8. Pour the popcorn into a bowl.

9. Sprinkle the hot popcorn with a pinch of salt, ⅛ cup of grated cheddar cheese, and/or 2 teaspoons of melted butter. Stir well with the spoon.

Results: How many cups of popcorn did you make? Compare your results to your prediction.

_____

_____

_____

_____

_____

CONTINUED

The Hows and Whys There are three important parts to every kernel of popcorn: a hard shell, water, and starch. When the popcorn kernels are heated, the water and starch both expand. The pressure of the expanding water forces the hard shell to crack open. When the starch gets free, it turns solid, making the fluffy popcorn that we love to eat. A lot of popcorn explodes out of just ¼ cup of kernels!

STEAM CONNECTION: Food scientists who create grocery store products like microwave popcorn and campfire popcorn run countless experiments trying to get the fluffiest, crunchiest popcorn possible.

Turn It Up! Try experimenting with different flavor combinations, like honey-cinnamon or garlic-sage. The easiest way to add a flavor to your popcorn is to melt 2 tablespoons of butter, mix in ½ to 1 teaspoon of spices (and 2 tablespoons of honey or sugar, if you're making a sweet flavor), and pour the butter-spice mixture over your freshly popped corn.

GODZILLA GUMMIES: OUTRAGEOUS OSMOSIS

LEVEL OF DIFFICULTY: EASY

MESS-O-METER RATING: MINOR MESS

BEST EATEN FOR: SNACK

PREP TIME: NONE

FULL TIME: 5 MINUTES BEFORE AND AFTER SOAKING OVERNIGHT

MAKES: 1 SNACK OF GUMMY BEARS

When a living thing is surrounded by a liquid that is very watery, we say that it is in a **hypotonic solution**. In this experiment, you will soak gummy bears in a variety of sugar solutions to test which ones are hypotonic. How will gummy bears change after they soak in different sugar solutions?

THE STEPS

1. Set 4 clear drinking glasses or jars on the counter.

2. Put a piece of tape on each glass. Label each glass: Water, slightly sugary water, very sugary water, and juice.

3. Pour 1 cup of water into each of the first three glasses.

4. Add ⅛ cup of sugar to the "slightly sugary water" glass and stir until the sugar dissolves.

5. Add ½ cup of sugar to the "very sugary water" glass and stir for 1 minute. Not all of the sugar will dissolve.

TOOLS:

- 4 clear glasses or jars
- Masking tape
- 1 spoon
- Ruler

INGREDIENTS:

- 3 cups water
- ⅛ cup sugar, plus ½ cup
- 1 cup juice
- 1 package gummy bears

Hypothesis: Predict the changes you will observe when you soak gummy bears in each liquid.

_____

_____

_____

_____

_____

_____

CONTINUED

Observations: Describe the changes you observed.

_____

_____

_____

_____

_____

_____

Results: Which solution made the gummy bear grow the biggest? Which is your most hypotonic solution? Why or why not?

_____

_____

_____

_____

_____

_____

_____

6. Add 1 cup of juice to the "juice" glass.

7. Put 1 gummy bear into each glass. Leave 1 gummy bear on the kitchen counter for comparison.

8. Put the glasses in the refrigerator and leave the gummy bears to soak overnight.

9. The next morning, use the spoon to take each gummy bear out of its glass. Use a ruler to measure each gummy bear from top to bottom. Record your observations and results.

10. All the gummy bears are safe to eat. Do they have different textures?

The Hows and Whys Each solution had more water and less sugar than the gummy bears. The water went into the gummy bears to balance out the amount of water and sugar between the bears and the solutions. The solution with the least sugar is the most hypotonic. The movement of water into (or out of) living things is called **osmosis**.

STEAM CONNECTION: Biologists study living things and their ability to survive in hypotonic solutions to learn more about the world around us.

Turn It Up! You can make hypotonic solutions with salt as well as sugar. Test a variety of foods, such as dried fruit and fruit snacks, in different hypotonic solutions.

IN A PICKLE: OVERNIGHT HORS D'OEUVRES

LEVEL OF DIFFICULTY: MEDIUM
MESS-O-METER RATING: MINOR MESS
BEST EATEN FOR: ON THE SIDE FOR LUNCH OR DINNER, SNACK
PREP TIME: 10 MINUTES TO SLICE THE CUCUMBERS AND PREPARE THE SYRUP
FULL TIME: 15 MINUTES BEFORE AND AFTER AN OVERNIGHT WAIT
MAKES: 1 QUART DILL PICKLES

TOOLS:

- Knife for an adult to slice cucumbers
- 1-quart glass jar with lid (or any other 1-quart heat-proof container with lid)
- 4-cup liquid measuring cup, heat-proof
- Measuring spoons
- Small pot with lid
- Stove
- Long-handled spoon
- Sieve

 When a living thing is surrounded by liquid that is very salty or sugary, we say that it is in a **hypertonic solution**. This can be dangerous for living things because they often lose water to the hypertonic solution. In this lab, you will make pickles. Pickles are made by soaking cucumbers in a salty, sugary hypertonic solution. How do you think the cucumbers will change?

Cautions: Have an adult complete steps 1, 4, and 5 (slicing the vegetables and handling the hot syrup).

THE STEPS

1. Have an adult use a knife to slice 4 pickling cucumbers into quarters the long way and ½ cup onion into thin

CONTINUED

slices, then use the broad side of the knife to smash 3 garlic cloves.

2. Put the sliced cucumbers and onions into a 1-quart jar. What do the cucumbers look like? Record your observations.

3. Use the 4-cup heat-proof liquid measuring cup to measure 1 cup of vinegar and 1 cup of water. Pour these into the pot. Add the 3 smashed garlic cloves, 3 teaspoons of salt, 3 teaspoons of dried dill, and either 1 tablespoon of sugar (for classic dill pickles) or ½ cup of sugar (for sweet dill pickles) to make the pickling syrup.

4. With adult supervision, heat the syrup on the stove over medium heat, stirring with a long-handled spoon at least once every minute, until it boils. Turn the heat off as soon as the syrup boils.

5. Have an adult pour the hot syrup into the 4-cup heat-proof liquid measuring cup. Note how many cups of syrup you have. Pour the hot syrup over the cucumbers and onion in the jar, then place the lid on the jar and put the jar immediately in the refrigerator.

6. Note how many cups of syrup are left in the measuring cup. Subtract this number from the total cups measured in step 5 to mathematically calculate how many cups of liquid you added to the pickles. Record your observations.

7. Let the jar sit in the refrigerator overnight.

8. The next day, place a sieve over the 4-cup heat-proof liquid measuring cup.

CONTINUED

INGREDIENTS:

- 4 pickling cucumbers (these are small and bumpy)
- ½ cup sliced onion
- 3 garlic cloves, smashed
- 1 cup vinegar
- 1 cup water
- 3 teaspoons salt
- 3 teaspoons dried dill
- 1 tablespoon (for classic dill pickles) to ½ cup sugar (for sweet dill pickles)

Hypothesis: Predict what changes you will observe after the cucumbers have soaked in the syrup overnight.

_____

_____

_____

_____

_____

_____

_____

Observations: Note the appearance of the cucumbers and the amount of liquid before (step 6) and after (step 9) the cucumbers soaked overnight.

_____
_____
_____
_____
_____
_____
_____
_____

Results: How did soaking overnight in a hypertonic solution change the cucumbers?

_____
_____
_____
_____
_____
_____
_____

9. Open the jar and pour everything into the sieve. The syrup will go into the measuring cup and your pickles will be in the sieve. How many cups of liquid are there? What do the pickles look like? Record your observations and results.

The Hows and Whys When the cucumbers were soaking in the syrup overnight, they lost a lot of water. This is because there was much more water in the cucumbers than in the syrup. The water went out of the cucumbers and into the syrup to balance out the amount of water between the cucumbers and the syrup. The movement of water out of (or into) living things is called osmosis.

STEAM CONNECTION: Biologists often grow living things in aquariums. They have to be careful to keep their aquarium water from getting hypertonic. If it does, the living things they study could lose water and get sick.

Turn It Up! Pickling is a food science technique that makes vegetables last longer without going bad. Try pickling other vegetables, like green beans or beets. Which vegetables lose the most water? The least?

TURNING PURPLE: HOMEMADE CABBAGE PH INDICATOR

LEVEL OF DIFFICULTY: MEDIUM

MESS-O-METER RATING: MINOR MESS

BEST EATEN FOR: LUNCH OR DINNER

PREP TIME: NONE

FULL TIME: 30 MINUTES

MAKES: 2 CUPS OF PURPLE VEGETABLE BROTH AND 2 CUPS OF PH INDICATOR

TOOLS:

- ➔ 2-quart pot with lid
- ➔ Measuring cups
- ➔ Stove
- ➔ Pint jar with lid
- ➔ Several small clear jars or glasses
- ➔ Several spoons

Acids are chemicals that taste sour and if they're very concentrated can burn through objects. Bases are chemicals that feel slippery and when concentrated can dissolve materials. When scientists want to know if something is an acid or a base, they use a technology called a **pH indicator**. This is a chemical that turns different colors when mixed with an acid or a base. In this experiment, you will cook up a pH indicator and use it to test beverages to discover if they are acids or bases. Do you think you have more acids or bases in your kitchen?

 **Cautions:** Adult supervision is needed while cooking on the stove.

CONTINUED

INGREDIENTS:

- ➡ ¼ head red cabbage
- ➡ 4 cups water, plus ¼ cup
- ➡ ¼ cup vinegar
- ➡ 1 tablespoon baking soda
- ➡ ¼ cup each of several beverages, such as milk, apple juice, lemon juice, and soda

Hypothesis: List the beverages that you will test. Next to each beverage, write whether you think it will be an acid or a base. For example, "Vinegar—acid."

_____

_____

_____

_____

_____

_____

_____

_____

THE STEPS

1. Put ¼ head of red cabbage and 4 cups of water in a 2-quart pot with a lid.

2. With adult supervision, put the lid on the pot and bring it to a boil over medium heat. Simmer the cabbage until the water turns a dark purple, 5 to 10 minutes.

3. Turn off the heat and let the broth cool for 10 minutes.

4. Have an adult carefully pour 2 cups of broth out of the pot and into the pint jar. This is your pH indicator. Set aside the pot, which will still have 2 cups of broth and the cabbage.

5. Pour ¼ cup of vinegar into one small jar.

6. Pour ¼ cup of water into a second small jar and stir in 1 tablespoon of baking soda.

7. Using a clean spoon, scoop 1 tablespoon of pH indicator out of its jar and add it to the vinegar. Repeat with the baking soda water. Record your observations.

8. Test the pH of several beverages by pouring ¼ cup of the beverage into a small jar and adding 1 tablespoon of your pH indicator. Record your results.

9. Put the lid on your jar of pH indicator and store it in the refrigerator. You will use this in the next two experiments, When Life Gives You Lemons (see page 163) and A Volcano in the Oven (see page 165).

10. Return to the pot that you set aside in step 4. You can drink your purple cabbage broth plain, or you can follow the *Turn It Up!* directions to make borscht.

The Hows and Whys Flavin, the molecule that makes cabbage broth purple, is sensitive to changes in pH. When flavin is in an acid solution, it goes through a chemical change and becomes a slightly different molecule that is pink. When flavin is in a base solution, it goes through a chemical change and becomes another molecule—one that is blue or green.

STEAM CONNECTION: Scientists often spend weeks refining the perfect recipes for the solutions they use. Many professional labs use solutions with over 20 important ingredients. Scientists use pH indicators like flavin to test the pH of their solutions.

Turn It Up! Borscht is a traditional Ukrainian food with beets, cabbages, and potatoes. To turn your purple cabbage broth into servings of borscht, begin with a large, empty soup pot. With adult supervision, heat 2 tablespoons of olive oil for 1 minute, then add 2 peeled, thinly sliced beets, 1 small chopped onion, and 1 stalk celery, sliced. Sauté the beets until they are tender, about 7 minutes. Add 2 cups of chicken broth, 2 cups of purple cabbage broth, 1 large cubed potato, and 1 large sliced carrot. While the soup is heating to a boil, add 1 bay leaf, 1 tablespoon of vinegar, and ½ teaspoon of salt. Simmer the soup until the vegetables are tender, about 10 minutes.Chop the cooked cabbage and add it to the hot soup. Serve with a dollop of plain yogurt or sour cream.

Observations: What color was the pH indicator when you first made it? What color did it turn when you added it to vinegar? How about when you added it to baking soda water?

_____

_____

_____

_____

_____

Results: List the beverages that you tested. Next to each beverage, write whether it turned the pH indicator pink (more acidic), blue (more basic), or stayed purple (neutral). Tally the number of acids and bases. Were there more acid beverages or more base beverages?

_____

_____

_____

_____

_____

WHEN LIFE GIVES YOU LEMONS:
THE PH OF LEMONADE

LEVEL OF DIFFICULTY: EASY

MESS-O-METER RATING: MINOR MESS

BEST EATEN FOR: A DRINK WITH LUNCH OR DINNER

PREP TIME: ABOUT 10 MINUTES TO JUICE THE LEMONS

FULL TIME: 30 MINUTES

MAKES: 4 CUPS OF LEMONADE

? In the experiment Turning Purple (see page 159), you made a pH indicator out of cabbage broth. In this experiment, you will use your pH indicator to compare the pH of lemon juice and lemonade. pH can be between 1 and 14. The lower the number, the more acid in the solution. When you test pH with cabbage broth, these colors will show you the pH.

| 1 | 2 | 3 | 4 | 5 | 6 | 7 | 8 | 9 | 10 | 11 | 12 | 13 | 14 |

← **Acidic** ——————— **Alkaline** →

When you add water to lemon juice to make lemonade, how much do you think the pH will change?

THE STEPS

1. Pour ⅛ cup of lemon juice into a small clear jar. Add 1 tablespoon of pH indicator. Record your observations.

CONTINUED

TOOLS:

- Citrus juicer
- 2 small clear jars or cups
- 4-cup liquid measuring cup
- Measuring spoons
- Spoon
- Drinking glasses

INGREDIENTS:

- Juice from 3 large lemons
- 3 cups water
- ½ cup sugar
- 2 tablespoons cabbage broth from Turning Purple (see page 159)

Hypothesis: Predict how much the pH of lemon juice will change when you mix in water to make lemonade.

_____

_____

_____

_____

Observations: What color did your pH indicator turn for the lemon juice? For the lemonade? Use the color chart at the start of this lab to find the pH of each liquid and write that down as well. For example: Lemon juice: purple: 6.

_____

_____

_____

_____

_____

_____

Results: Subtract the pH of the lemon juice from the pH of the lemonade to calculate the change in pH.

_____

_____

_____

_____

_____

2. Pour 3 cups of water into the 4-cup liquid measuring cup. Add lemon juice until the total liquid measures 3½ cups. Add ½ cup of sugar.

3. Stir everything together well with the spoon.

4. Pour ⅛ cup of lemonade into the second small clear jar. Add 1 tablespoon of purple cabbage broth pH indicator. Record your observations and results.

5. Enjoy the remaining lemonade plain or with ice!

The Hows and Whys To make a major shift in the pH of a solution, a chemical change is needed. Adding water to lemon juice dilutes the acid a little bit but doesn't make a chemical change. Making lemonade does not change the pH of lemon juice very much.

STEAM CONNECTION: Scientists add water to solutions to dilute them. Biologists dilute solutions several times to make a culture of tiny organisms less crowded. Chemists dilute solutions to get the right concentration of molecules for their experiments.

Turn It Up! Now that you know that you need to add 10 times as much water to raise the pH of an acid by 1, design an experiment to dilute lemon juice enough to raise its pH by 2. Be sure to include measurements in your plan.

A VOLCANO IN THE OVEN: BAKING SODA & VINEGAR CAKE

LEVEL OF DIFFICULTY: ADVANCED

MESS-O-METER RATING: MINOR MESS

BEST EATEN FOR: DESSERT

PREP TIME: SET THE BUTTER OUT OF THE REFRIGERATOR 2 HOURS BEFORE BEGINNING

FULL TIME: 30 MINUTES TO MAKE THE BATTER, 15 MINUTES TO BAKE, AND 5 MINUTES TO COOL

MAKES: 24 CUPCAKES

TOOLS:

- Oven
- Oven mitts
- 2 muffin tins
- 24 paper or silicone muffin tin liners
- 2 large bowls
- Measuring cups and spoons
- Electric mixer, stand mixer with paddle attachment, or spoon
- Large spoon
- 3 small clear jars or glasses
- 2 regular spoons

The classic volcano experiment combines vinegar (an acid) with baking soda (a base) to model a volcanic eruption. In this experiment, you will use the pH indicator from Turning Purple (see page 159) to compare the pH of vinegar with the pH of cupcake batter. pH can be any number between 1 and 14. The lower the number, the more acid in the solution. The chart on page 163 will show you the colors you will see for each pH.

When you add vinegar to baking soda and other ingredients to make cupcake batter, how much do you think the pH will change?

 **Cautions:** Adult supervision is needed while using any electric mixers and the oven.

CONTINUED

INGREDIENTS:

- ➔ ¾ cup unsalted butter, room temperature
- ➔ 1¾ cups sugar
- ➔ 1 egg
- ➔ 2½ cups flour
- ➔ 1¼ teaspoons baking soda
- ➔ 1¼ teaspoons cinnamon
- ➔ ½ teaspoon salt
- ➔ ½ cup cocoa powder
- ➔ 1½ cups water
- ➔ 1 teaspoon vanilla
- ➔ 3 tablespoons vinegar, divided
- ➔ 3 tablespoons purple cabbage broth from previous experiment Turning Purple (see page 159), divided

THE STEPS

1. Preheat the oven to 350°F. Place 1 paper or silicone baking cup in each cup of the muffin tin.

2. Combine ¾ cup of butter, 1¾ cups of sugar, and 1 egg in one large bowl. With adult supervision, use an electric mixer, the paddle attachment on a stand mixer, or a spoon to stir everything up well.

3. In another large bowl, combine 2½ cups of flour, 1¼ teaspoons of baking soda, 1¼ teaspoons of cinnamon, ½ teaspoon of salt, and ½ cup of cocoa powder. Use a clean, dry spoon to stir the dry ingredients until the mixture is evenly light brown in color.

4. Add the dry ingredients to the butter, sugar, and eggs. With adult supervision, use an electric mixer, the paddle attachment on a stand mixer, or a spoon to stir everything well.

5. Add 1½ cups of water and 1 teaspoon of vanilla to the batter. With adult supervision, use an electric mixer, the paddle attachment on a stand mixer, or a spoon to stir everything well.

6. Pour 1 tablespoon of vinegar into a small glass jar. Add 1 tablespoon of purple cabbage broth pH indicator. Record your observations.

7. Use a regular spoon to scoop 1 tablespoon of cupcake batter out of the mixing bowl and place the spoon into a small glass jar. Add 1 tablespoon of purple cabbage broth pH indicator. Gently stir the mixture with the spoon. Allow the clouds of batter to settle down so

you can see the color of the pH indicator. Record your observations.

8. Add 2 tablespoons of vinegar to the large bowl of cupcake batter. With adult supervision, use an electric mixer, the paddle attachment on a stand mixer, or a spoon to stir everything well.

9. Use a regular spoon to scoop 1 tablespoon of cupcake batter with vinegar out of the mixing bowl and place the spoon into a small glass jar. Add 1 tablespoon of purple cabbage broth pH indicator. Gently stir the mixture with the spoon. Allow the clouds of batter to settle down so you can see the color of the pH indicator. Record your observations.

10. Divide the rest of the cupcake batter evenly between the 24 cups of the 2 muffin tins. Each cup will be ⅔ to ¾ full.

11. With adult supervision, place the muffin tins in the preheated 350°F oven. Bake 15 minutes. When the cupcakes are done, a toothpick inserted into a cupcake will come out clean.

12. While the cupcakes are baking, record your results.

13. These cupcakes are delicious plain, but you can also frost them with frosting from the Over the Rainbow experiment (see page 122) in chapter 6.

Hypothesis: Predict how much the pH of vinegar will change when you mix it into the cupcake batter. For example, if you expect the pH to go from 4 to 6, the change would be 2.

_____

_____

_____

_____

_____

Observations: What color did your pH indicator turn for the vinegar? For the cupcake batter without vinegar? For the cupcake batter with vinegar? Use the color chart on page 163 to find the pH of each liquid and write that down as well. For example: Vinegar: purple: 6.

_____

_____

_____

_____

_____

CONTINUED

Results: Subtract the pH of the vinegar from the pH of the cupcake batter to calculate the change in pH.

_____

_____

_____

_____

_____

The Hows and Whys In this chemical reaction, the vinegar and baking soda combine to make new chemicals. The acid disappears from the solution because it goes through a chemical change. Bubbles were made by the chemical reaction between the vinegar and baking soda. These cupcakes have a fluffy texture because of the bubbles.

STEAM CONNECTION: When chemists need to make a solution less acidic or basic, they will add a chemical that changes the pH of the solution. This is especially useful when chemists have chemicals that would hurt the environment if their pH were not changed. Putting strong acids into the environment can create acid rain, which is harmful to living things.

Turn It Up! Experiment with the reaction between vinegar and baking soda by combining measured amounts of the two chemicals in a drinking glass. Eventually, you will figure out the right proportion so that you can gently overflow the glass. If you add red food coloring to the vinegar before the reaction and set the glass inside a volcano model, you will have a classic model volcano that erupts when you add baking soda!

DEEP FREEZE:
ICE CREAM ON A ROLL

LEVEL OF DIFFICULTY: EASY
MESS-O-METER RATING: MEDIUM MESS
BEST EATEN FOR: DESSERT
PREP TIME: NONE
FULL TIME: 30 MINUTES
MAKES: 4 SERVINGS OF SOFT VANILLA ICE CREAM

TOOLS:

- 2-cup liquid measuring cup
- Measuring cups and spoons
- 2-cup jar with a secure lid
- 3-pound coffee can
- Duct tape (about 3 feet)
- Weather thermometer

INGREDIENTS:

- 1 cup milk
- ¾ cup cream
- ⅓ cup sugar
- ½ teaspoon vanilla
- 3 to 6 trays ice cubes
- 1 cup ice cream salt

 In the winter, people often sprinkle salt on sidewalks and roads because it can melt ice. In this experiment, you will explore the effect of salt on the temperature of ice and use the effect to make soft serve ice cream. Regular ice is 32°F. How do you think salt will change the temperature of ice?

Cautions: Have an adult check your lids to make sure they are secure. Adult supervision is needed when working with duct tape.

CONTINUED

Hypothesis: Predict the temperature of ice when it is mixed with salt.

_____

_____

_____

_____

_____

Observations: Record the temperature of the ice before and after you roll or kick the can.

_____

_____

_____

_____

_____

Results: By how many degrees did the temperature of the ice change? Subtract the final temperature from the starting temperature. Did the salt raise or lower the temperature of the ice?

_____

_____

_____

_____

_____

THE STEPS

1. Pour 1 cup of milk into a 2-cup liquid measuring cup.

2. Add cream on top of the milk to bring the total liquid to 1¾ cups. How much cream did you add?

3. Pour ⅓ cup of sugar into a 2-cup jar with a secure lid.

4. Pour as much of the milk and cream mixture as will fit into the jar.

5. Pour ½ teaspoon of vanilla into the jar.

6. Have an adult help you get the lid firmly on the jar.

7. Shake the jar well so that the ingredients blend together.

8. Place the jar with the ice cream solution inside a 3-pound coffee can.

9. Pour ice over the ice cream jar so that it fills the space between the jar and the coffee can.

10. Add 1 cup of ice cream salt on top of the ice.

11. Have an adult help you get the lid firmly on the coffee can. You might want to use duct tape to secure the lid to the can.

12. Gently roll the can back and forth for 15 minutes. You can do this on the kitchen counter, on the kitchen floor, or on the sidewalk outside. You can use your hands or your feet, but be gentle. You do not want the can to break open.

13. After 15 minutes of rolling, open the coffee can. Use a thermometer to take the temperature of the ice and salt inside the coffee can. Record your observations.

14. Ice is normally 32°F. Subtract the temperature of your ice and salt to mathematically calculate the number of degrees that the salt cooled the ice. Record your results.

15. Enjoy your ice cream!

The Hows and Whys When ice melts, it uses up heat. The water molecules cool down when ice melts because it takes energy for ice to melt into water. When salt is added to ice water, the water can't refreeze because the salt molecules get in the way—but it gets colder and colder (even as low as -6°F!) as more ice melts.

STEAM CONNECTION: The physical changes of freezing and melting are very important to Earth scientists who study the atmosphere and weather.

Turn It Up! You can change the temperature at which water freezes and melts by adding salt or other ingredients, like sugar. Were you satisfied with the texture of your ice cream? Do you want it to freeze harder? Go online to discover whether there are any tasty ingredients you could add to your ice cream solution to raise its freezing temperature and get it to freeze harder—or experiment with the amount of ice added to your ice cream maker to see if you can get your ice even colder.

SLIME-Y SCIENCE:
GOO YOU CAN CHEW

LEVEL OF DIFFICULTY: EASY
MESS-O-METER RATING: MEDIUM MESS
BEST EATEN FOR: SNACK, DESSERT
PREP TIME: 5 MINUTES TO HELP MEASURE OUT THE MARSHMALLOW FLUFF
FULL TIME: UP TO 30 MINUTES
MAKES: 1½ CUPS EDIBLE MARSHMALLOW SLIME

TOOLS:

- Bowl
- Measuring cups and spoons
- Spoon
- Plastic storage container with lid

INGREDIENTS:

- 1 cup marshmallow fluff
- ½ teaspoon vanilla
- ¼ cup powdered sugar
- 1 cup cornstarch, divided
- Gel or regular food color (optional)

 Different ingredients bring different textures to the foods we prepare. Often, the relative amounts, or **ratio**, of each ingredient makes a big difference. For example, cake has a flour-to-egg ratio of 1:1 (1 cup of flour for 1 cup of eggs), while pancakes have a flour-to-egg ratio of 2:1 (2 cups of flour for 1 cup of eggs). In this experiment, you will try different ratios of marshmallow fluff to cornstarch to make slime with the perfect texture. What ratio of marshmallow fluff to cornstarch will make slime that is stretchy but not sticky?

Cautions: Eating too much uncooked cornstarch will upset your stomach. Although this slime is safe to taste, do not eat all of it at once.

THE STEPS

1. Pour 1 cup of marshmallow fluff, ½ teaspoon of vanilla, ¼ cup of powdered sugar, and ¼ cup of cornstarch into

CONTINUED

Hypothesis: Predict the ratio of marshmallow fluff to cornstarch that will make slime that is stretchy but not sticky. For example, 3 parts marshmallow to 1 part cornstarch.

_____

_____

_____

_____

_____

Observations: Describe the texture of your slime at each stage of the experiment.

_____

_____

_____

_____

_____

Results: What was the final ratio of marshmallow fluff to cornstarch?

_____

_____

_____

_____

_____

a bowl. This mixture has a ratio of 4 parts marshmallow fluff to 1 part cornstarch.

2. Stir the mixture with a spoon until everything sticks together. Record your observations.

3. Add another ¼ cup of cornstarch into the bowl. Stir the mixture with a spoon until everything sticks together. You now have 1 cup of marshmallow fluff and ½ cup of cornstarch. What is the ratio of this mixture? Record your observations.

4. Add another ¼ cup of cornstarch into the bowl. Stir the mixture with a spoon until everything sticks together. What is the ratio of this mixture? Record your observations.

5. Pick up your ball of slime with clean hands. Play with the slime for a few minutes.

6. If you like the texture of your slime, record your results. If it is too sticky, add another ¼ cup of cornstarch into the bowl. Stir the mixture with a spoon until everything sticks together. What is the ratio of this mixture? Record your observations and results.

7. If you want color slime, add 3 to 5 drops of food color and work it into the ball of slime with your hands.

8. Store your slime in a plastic storage container with a sprinkling of cornstarch and the lid on tight.

The Hows and Whys Cornstarch is a thickening agent that traps liquids and makes solutions less sticky. When you added cornstarch to marshmallow fluff, it trapped the moisture and made it less sticky. A 1:1 ratio of marshmallow fluff to cornstarch usually works, but when you handle your slime, it heats up. This can wear out the cornstarch. You may need to add more cornstarch each time you handle your slime.

STEAM CONNECTION: Food scientists have to work with ratios to get recipes right. In a food lab, scientists will often cook the same recipe with slightly different ratios—over and over again—until they get it right.

Turn It Up! You can also make marshmallow slime with whole marshmallows. You will need 1 cup of mini marshmallows instead of 1 cup of marshmallow fluff. Follow the same recipe but have an adult heat the mini marshmallows for about 20 seconds in the microwave to make them soft. Try adding mix-ins to your marshmallow slime. What would happen if you added chocolate chips?

IT'S A GAS: FLUFFY BAKED PANCAKE

LEVEL OF DIFFICULTY: ADVANCED
MESS-O-METER RATING: MINOR MESS
BEST EATEN FOR: BREAKFAST
PREP TIME: NONE
FULL TIME: 45 MINUTES
MAKES: 1 LARGE FLUFFY PANCAKE

 When a gas gets hot, it gets bigger and takes up more volume. Bakers use this law of physics whenever they make something fluffy in the oven. In this experiment, you will bake a fluffy pancake out of eggs, flour, and milk. By what factor will your baked pancake grow?

Cautions: Have an adult handle the oven and hot skillet using oven mitts.

TOOLS:

- 8- to 10-inch ovenproof skillet
- Oven
- 2-cup liquid measuring cup
- Measuring cups
- Fork
- Oven mitts
- Ruler

INGREDIENTS:

- 2 tablespoons butter
- ½ cup milk
- 3 eggs
- ½ cup flour
- Maple syrup to taste

THE STEPS

1. Put 2 tablespoons of butter in an ovenproof skillet.

2. Place the skillet in a cold oven.

3. Preheat the oven to 425°F.

4. Pour ½ cup of milk into a 2-cup liquid measuring cup.

5. Crack 3 eggs into the milk and stir with a fork.

6. Use the fork to beat ½ cup of flour into the egg and milk mixture.

7. Have an adult use oven mitts to take the hot skillet out of the oven and pour the batter into the hot skillet. Estimate the height of the batter in the skillet by holding a ruler outside of the skillet. Do not touch the hot skillet. Record your observations.

8. Return the skillet with the batter in it to the oven and bake at 425°F for 25 minutes. The pancake will look puffy and golden-brown when it is finished baking.

9. Have an adult use oven mitts to remove the cooked pancake from the oven.

10. Use a ruler to measure the height of the pancake before it collapses. Record your observations and results.

CONTINUED

Hypothesis: Predict the factor by which your pancake will grow when it gets hot in the oven. For example, if your pancake starts out at 1 inch and grows to 3 inches, it will have grown by a factor of 3.

_____

_____

_____

_____

_____

Observations: Record the height of your pancake before and after baking.

_____

_____

_____

_____

_____

Results: Divide the height of your baked pancake by the height of your batter to calculate the factor by which it grew.

_____
_____
_____
_____
_____

The Hows and Whys When your pancake was cooking, the water molecules from the eggs and milk changed from liquid to gas. The volume of the gas got bigger and bigger as the pancake got hotter and hotter. The gas couldn't escape because the molecules of flour trapped it inside. So the pancake had to get bigger to make space for that huge volume of gas. My pancakes usually grow by a factor of 3 or 4.

STEAM CONNECTION: Food scientists make use of the expansion of heated gas when they make pancakes, muffins, soufflés, and meringues. Engineers need to know how much heated gases will expand when they work with hot air balloons, scuba equipment, or compressed air tanks.

Turn It Up! The flavor possibilities with this recipe are incredible! Try adding ½ cup of chopped spinach to the batter, then sprinkling ½ cup of grated cheddar cheese over the top for the last 5 minutes of baking time. Or mix in ½ cup of grated apples and top with ⅛ cup of sugar mixed with ½ teaspoon of cinnamon. What flavor combinations would you enjoy?

EXTRA FEATURE: THE SCIENCE OF NUTRITION

Eating healthy feels great and is easy—if you know which foods are healthy. The science of nutrition explores which foods help people be healthy and which foods to avoid. In the 1700s, nutrition scientists figured out which vitamins and minerals people needed to eat to avoid certain diseases. For example, vitamin C prevents and cures scurvy, a disease that sailors got when they didn't eat enough fruits and vegetables.

In modern times, most people eat a variety of foods and aren't missing important vitamins or minerals. Instead, people have to worry about eating too many unhealthy foods. Nutrition scientists work to discover the right balance of foods to recommend for healthy eating. Here are a few general food rules that work for just about everybody:

- Fruits and vegetables are good for your body.
- Whole grains are healthier than processed grains. For example, whole wheat bread is healthier than white bread and brown rice is healthier than white rice.
- Lean proteins—like fish, nuts, peas, beans, and low-fat milk, cheese, and yogurt—are good for you.
- It's best to limit salty, sugary, and high-fat foods and drinks.
- Whole foods (like almonds) are usually healthier than processed foods (like almond-flavored crackers).

Each person needs to eat a little differently to feel their best. Check out the MyPlate website in the Resources to find out more. And remember, you can't go wrong with a fresh fruit or vegetable!

Chapter Eight

IT'S A WRAP

Your kitchen experiments activated all the parts of STEAM: asking scientific questions, developing cool technologies, designing engineering solutions, flexing your artistic creativity, and solving sweet mathematical calculations. In your kitchen, as in professional labs, the branches of science worked together. You've explored the science of food and used food to explore science.

When you understand the science behind food, you can cook food that is healthier and more delicious. Now that you know how to blanch and cold shock your vegetables, you'll never have to eat a mushy carrot again. Now that you know how to make a dynamo salad dressing, you can enjoy every crunchy bite of lettuce. Learn how to make your least favorite food your most favorite food by bringing its hidden flavors to life.

As you cook, keep investigating the questions that pop up. When you see something strange or interesting happen in your pan, write down a question about it. The food science books and websites in the Resources section will help answer your questions—and you can always design your own experiment to find out more. Also check out the suggested cookbooks and websites. There is a whole world of amazing recipes out there, waiting for you to cook them—and eat them!

RESOURCES

STORYBOOKS

Walter the Baker by Eric Carle
Learn about the difference between yeast bread made with milk or water in this historic book for young children.

The Giant Jam Sandwich by John Vernon Lord
A funny story about how a village fought an invasion of wasps by baking a huge loaf of bread.

Strega Nona by Tomie dePaola
Big Anthony runs into trouble when he tries to make pasta with his employer's magic pasta pot.

The Little Red Hen Makes a Pizza by Philomen Sturges and Amy Walrod
In this modern twist on the classic fable, the Little Red Hen makes an enormous pizza—step by step—then shares it with her non-chef friends.

FOOD SCIENCE REFERENCE BOOKS

What Einstein Told His Cook and What Einstein Told His Cook 2 by Robert L. Wolke
Two books explaining the science of everyday cooking through short but advanced articles that would make great read-alouds.

The Food Lab by J. Kenji López-Alt and The Science of Good Cooking from Cook's Illustrated
Two books with fascinating background reading for adults curious about exploring food science in more detail.

COOKBOOKS

Butter Flour Sugar Eggs: Whimsical Irresistible Desserts by Gale Gand, Rick Tramonto, and Julia Moskin
My go-to baking cookbook, with amazing recipes for whole wheat banana bread, caramel rolls, and raspberry-stuffed French toast.

Martha Stewart's Cookies: The Very Best Treats to Bake and Share: A Baking Book by Martha Stewart Living magazine

When I get ready to make a batch of cookies, I start with this book. The table of contents is a series of cookie photos, which helps younger children select the yummiest looking cookies to bake. My gingerbread recipe is inspired by this book.

Joy of Cooking by Irma S. Rombauer, Marion Rombuer Becker, and Ethan Becker

With detailed articles on hundreds of cooking techniques and over 4,000 recipes, the *Joy of Cooking* can answer almost any cooking question. If you're not sure how to bake a squash or want to bake a basic yellow cake from scratch, this is your book. Mine is covered in splashes and stains and crammed full of bookmarks.

Angry Trout Café Notebook: Friends, Recipes, and the Culture of Sustainability by George Wilkes

The Angry Trout is a marvelous restaurant at the northern edge of Minnesota. Their cookbook has the best salad dressing recipes in the world, plus a treasure trove of ideas for cooking fish and making chowders. My salad dressing recipe is inspired by theirs.

WEBSITES

Choose My Plate (www.choosemyplate.gov)
Guidelines for healthy eating from the American government, with a great children's website full of games and activities.

Scientific American (www.scientificamerican.com/education/bring-science-home/)
Easy science experiments for kids ages 6 to 12 to try at home from *Scientific American* magazine.

Steam Powered Family (www.steampoweredfamily.com)
Tons of free science experiment ideas, including 40 edible science experiments.

Steve Spangler Science (www.stevespanglerscience.com)
Fun, easy STEAM experiments with great video tutorials.

Coffee Cups and Crayons (www.coffeecupsandcrayons.com)
DIY kitchen science, edible art, and experiments.

We Are Teachers (www.weareteachers.com/edible-science/)
Advanced edible experiments, including atom and DNA models.

Exploratorium (www.exploratorium.edu /cooking/candy/sugar-stages.html)
An explanation of the incredible variety of sugar candy stages.

The Kitchn (www.thekitchn.com)
Articles and recipes exploring foods and food science.

Epicurious (www.epicurious.com)
Searchable database of every recipe ever published in *Gourmet* and *Bon Appétit* magazines, including a Chocolate Raspberry Soufflé recipe by Katherine Sacks.

BritCo (www.brit.co/flavored-butter/)
Flavored Butter Recipes from BritCo.

Realtor Magazine (magazine.realtor/home -and-design/guide-residential-styles)
Realtor Magazine's guide to residential styles of houses, which could provide inspiration for your gingerbread house.

Smitten Kitchen (smittenkitchen.com)
Cooking blog with some of the world's most delicious recipes.

ORGANIZATIONS

Institute of Food Science + Technology (www.ifst.org/lovefoodlovescience /resources)
Over three dozen advanced food science experiments and other resources from Love Food Love Science by the Institute of Food Science + Technology.

The American Chemical Society (www.acs.org)
The American Chemical Society, a group of experienced chemistry teachers, shares oodles of free lab ideas, many of which involve food science, on its website.

Science Buddies (www.sciencebuddies.org)
Tons of experiments you can do at home, including a river model using cornmeal, sand, and water, from Science Buddies.

GLOSSARY

ABSORB: to soak up a liquid

ACID: a sour-tasting chemical that burns through materials

AGRICULTURE: the science of growing food

ANALYZE: to study something carefully to understand what it means

ARCHIMEDES' PRINCIPLE: an object in water is lifted up by a force equal to the weight of the water that is moved by the object

ARCHITECTURE: the science of designing and making buildings

ARTIFICIAL: made by humans, not nature

ARTS: creative works, including painting, drawing, writing, dance, music, and theater

BACTERIA: tiny, single-celled living things

BASE: a slippery chemical that dissolves materials

BEDROCK: the solid rock that's under Earth's surface dirt

BIOLOGY: the science of life

BIOLUMINESCENCE: when living things give off light

BIOTECHNOLOGY: technology used with or for living things

BLANCHING: boiling for a short period of time

CALORIE: a measure of energy in food

CALORIMETER: a machine that calculates how many calories are in a serving of food

CAPILLARY ACTION: the movement of water up or down a tube

CARBONATION: the fizz in drinks, which comes from carbon dioxide gas in the liquid

CAST: a fossil formed when rock replaces the remains of a living thing

CATALYZE: to speed up a chemical reaction

CHEMICAL CHANGE: when the atoms in a molecule change or move, making a new chemical

CHEMICAL ENGINEERING: engineering that applies to chemicals

CHEMICAL REACTION: when molecule(s) change

CHEMISTRY: the science of matter and how it reacts and can change

CHLOROPHYLL: the molecule in plants that captures sunlight energy to make sugars

CHROMATOGRAPHY: technology for separating a mixture using paper or other materials

COLD SHOCK: plunging freshly cooked foods into ice water to stop the cooking process

COLOR WHEEL: a circle showing the relationships between colors

COMBUSTION: a chemical change involving burning

CONDUCT: heat movement through a substance

CONTROL: a version of an experiment where everything is identical to the experiment except the one variable that is being tested

CRYSTALS: a solid with regular shapes at the molecular level

CULINARY: related to food or cooking

CULTURE: the controlled growth of bacteria or other tiny living things

CURDS: the milk solids that turn into cheese

DENATURE: to undo the shape of a molecule

DENSITY: the ratio of mass to volume

DENSITY COLUMN: a cylinder filled with layers of liquids of different densities

DESIGN PROCESS: steps for solving problems in engineering

DISSECT: to take apart

DNA: the molecule that codes for genes in living things and is different for every organism

EARTH SCIENCE: the study of Earth's rocks and weather

EFFICIENT: maximum results in minimum time

ENGINE: a machine that uses fuel to move

ENGINEERING: the science of designing, building, and solving problems

ENZYME: a type of protein that catalyzes chemical changes

EXTRACT: to remove

FLAVIN: the molecule that makes purple cabbage broth purple

FOSSILS: the preserved remains of living things from long ago

FUNGUS: a living thing that grows from spores, including yeast, mushrooms, and molds

GARNISH: edible decoration

GAS: matter with molecules that can spread far apart to fill any container

GEOMETRY: the math of shapes

GLACIERS: big, heavy mountains of ice

GLOBAL CLIMATE CHANGE: human-caused changes to Earth's overall climate

GRAVITY: a force that pulls two objects together, especially toward Earth

HYPERTONIC SOLUTION: a solution with a high concentration of salts, sugars, or other dissolved substances

HYPOTHESIS: a scientific prediction

HYPOTONIC SOLUTION: a solution with a low concentration of salts, sugars, or other dissolved substances

ICE AGE: a cold time in history when glaciers covered a lot of the Earth, from 2.4 million years ago until 11,000 years ago

IGNEOUS: rock that is made of cooled lava or magma

LAVA: liquid rock above ground

LEAVENING: something that makes batter or dough rise when it cooks

LEVER: a stick that helps move a heavy load

LIQUID: matter with molecules that slide past each other and flow

MAGMA: liquid rock below ground

MASS: a measurement of how much matter is in an object (on Earth, this is the same as an object's weight)

MATHEMATICS: the language of numbers

MATTER: any physical material

MECHANICAL ENGINEERING: engineering for machines

METAMORPHIC: rock that formed under high heat or pressure

MICROBIOLOGY: the study of tiny organisms

MIXTURE: two or more things that are combined but can be easily separated because they have not had a chemical change

MODELING: a smaller or simpler version of a large or complicated process or thing

MOLDS: hollows in rock from the remains of living things that dissolved away

MOLECULAR GASTRONOMY: food science focused on physical and chemical changes

MOLECULE: a group of atoms bonded in a certain shape and proportion

MORDANT: a chemical that helps dye work

NATURAL: from nature, not human-made

OPTICS: the physics of light

ORGANISM: an individual living thing

OSMOSIS: the movement of water into or out of living things

PECTIN: a molecule in fruit that makes jam set

PERIODIC TABLE: a chart organizing all the elements in the universe according to their physical and chemical properties

PH: a number showing the concentration of acid, or H_3O^+ molecules, in a solution

PH INDICATOR: a chemical that changes colors when mixed with an acid or a base

PHASE CHANGE: a change from solid to liquid to gas

PHOTOSYNTHESIS: the process that plants use to make food from sunlight energy, water, and air

PHYSICAL CHANGE: a change in the form of a chemical, but not its molecules

PHYSICAL PROPERTY: trait of a chemical that does not involve chemical change, such as boiling point, density, or texture

PHYSICS: the science of matter and energy

PIGMENT: a molecule that has a color

PITH: the white stuff inside citrus fruit

POLYHEDRON: a 3-dimensional object made of many flat shapes

PRESERVED REMAINS: fossils that are still whole

PRIMARY COLOR: a color that can be used to make another color by mixing

PROTEIN: a large molecule made of chains of amino acids

PROTOTYPE: a first model of something one plans to build

PSYCHOLOGIST: a scientist who studies the human mind

QUESO FRESCO: a simple, fresh cheese

RADIATION: energy that comes out of something, like light and heat from the Sun

RATIO: relative amounts of two or more ingredients

RESULTS: what you discover in an experiment

SCIENCE: learning about the world by making observations, thinking about data, and noticing patterns

SCIENTIFIC METHOD: a series of steps that can be used to answer a scientific question

SECONDARY COLOR: a color that can be made from other colors by mixing

SEDIMENTARY: rock that formed when loose dirt, sand, and rock piled up

SOLID: matter with molecules that stay in place

SOLUTION: a liquid with one or more molecules evenly dissolved inside

SPHERIFICATION: turning something (like a liquid) into a sphere

STARCH: a large molecule made of chains of sugars

TECHNOLOGY: science that is applied to solve problems

THICKENING AGENTS: chemicals added to foods to make them less liquid

TRACE FOSSILS: fossils created by living things, without any remains, like a footprint

VOLUME: how much space a thing takes up

WHEY: the liquid left over in cheese-making

INDEX

ACKNOWLEDGMENTS

Ten thousand thank-yous to Orli Zuravicky and the Callisto staff for dreaming up this fantastic book. I will always be grateful to you for choosing me to write it and for bringing it to life with your creativity and expertise.

Thank you to my loving kids, for trying out the recipes; to my forbearing husband, for washing up after 50 kitchen experiments; and to all the strong women in my family who taught me to find joy in the kitchen: Granny, whose encyclopedic knowledge of American cuisine had our family in raptures for 94 beautiful years; the Curnow clan, whose enthusiasm for baking has firmly established pie as the only acceptable breakfast food the day after a family gathering; and my mother, who raised me on a diet of wildly experimental and unavoidably healthy foods.

ABOUT THE AUTHOR

Dr. Megan Olivia Hall is the 2013 Minnesota Teacher of the Year and the 2015 Minnesota Teacher of Excellence. In her 20 years of teaching science, Megan has worked with learners of many ages and levels, from kindergarteners to graduate students. A National Board Certified Teacher, she serves as science department chair and develops curriculum at Open World Learning Community in St. Paul Public Schools. Megan's writing has been featured in *Education Week* and *The Science Teacher*. A Leading Educator Ambassador for Equity Fellow with the Education Civil Rights Alliance, Megan holds a PhD in Learning, Instruction, and Innovation from Walden University. She lives in St. Paul with her husband, two kids, two cats, and a messy, edible garden.

NOTES